Methods in Molecular Biology

Series Editor
John M. Walker
School of Life and Medical Sciences
University of Hertfordshire
Hatfield, Hertfordshire, UK

For further volumes:
http://www.springer.com/series/7651

For over 35 years, biological scientists have come to rely on the research protocols and methodologies in the critically acclaimed *Methods in Molecular Biology* series. The series was the first to introduce the step-by-step protocols approach that has become the standard in all biomedical protocol publishing. Each protocol is provided in readily-reproducible step-by-step fashion, opening with an introductory overview, a list of the materials and reagents needed to complete the experiment, and followed by a detailed procedure that is supported with a helpful notes section offering tips and tricks of the trade as well as troubleshooting advice. These hallmark features were introduced by series editor Dr. John Walker and constitute the key ingredient in each and every volume of the *Methods in Molecular Biology* series. Tested and trusted, comprehensive and reliable, all protocols from the series are indexed in PubMed.

Dynamin Superfamily GTPases

Methods and Protocols

Edited by

Rajesh Ramachandran

School of Medicine, Case Western Reserve University, Cleveland, OH, USA

Editor
Rajesh Ramachandran
School of Medicine
Case Western Reserve University
Cleveland, OH, USA

ISSN 1064-3745 ISSN 1940-6029 (electronic)
Methods in Molecular Biology
ISBN 978-1-0716-0675-9 ISBN 978-1-0716-0676-6 (eBook)
https://doi.org/10.1007/978-1-0716-0676-6

This Humana imprint is published by the registered company Springer Science+Business Media, LLC, part of Springer Nature.
The registered company address is: 1 New York Plaza, New York, NY 10004, U.S.A.

Preface

Dynamin superfamily proteins or DSPs comprise a diverse collection of large, self-assembling, mechanoenzymatic GTPases that catalyze various intracellular membrane remodeling events ranging from the scission of highly curved transport vesicles from relatively flat parent membranes (e.g., as in endocytosis) to the fission and fusion of large organelles such as the mitochondria and chloroplasts. Although DSPs are commonly found from bacteria to man, their roles, mechanisms, and significance remain disproportionately better characterized in the baker's yeast (*Saccharomyces cerevisiae*) and in mammals than in bacteria or plants.

DSPs may be roughly classified into *ancient* and *modern* dynamins. *Modern* dynamins, found exclusively in higher eukaryotes with a well-defined nervous system, include the three isoforms of dynamin (Dyn1-3), the prototypical, founding member of this protein superfamily. These *modern* dynamins are distinguished from their *ancient* but extant counterparts by the presence of a canonical pleckstrin homology (PH) domain and an unstructured proline-rich domain (PRD). In their place, however, *ancient* dynamins conserve a veritable collection of largely disordered inserts, ranging in length from a few to over a hundred residues that albeit perform corollary roles in selective protein-lipid and protein-protein interactions, respectively.

Unicellular yeast (and other fungi) encodes for only four DSPs, all *ancient*, namely Vps1, Dnm1, Mgm1, and Fzo1. Whereas Vps1, which functions in various intracellular membrane remodeling events in yeast, has no known counterpart in higher eukaryotes, the other three possess orthologs in dynamin-related protein 1 (Drp1), mitofusins (Mfn), and optic atrophy 1 (OPA1), all of which participate in mitochondrial dynamics—the regulated catalysis of cyclical mitochondrial double membrane fission and fusion. Either coincident with or apart from their roles in the regulation of membrane dynamics, DSPs such as Mx (myxovirus resistance) and hGBP1 (human guanylate binding protein 1) found in higher eukaryotes also confer resistance to viral infection (e.g., HIV), albeit through less well-understood mechanisms. Among other commonalities, *modern* and *ancient* dynamins both function in the remodeling of the actin cytoskeleton either in concert with or independently of membranes, although this aspect needs further elucidation. Regardless, it is now widely appreciated that DSPs own a broader array of functions than originally posited.

Much of what we know about DSP mechanism(s) in membrane dynamics is largely owed to the bottom-up reconstitution of its function using isolated protein molecules and biomimetic synthetic lipid bilayers that closely approximate the topology of the native, target membrane. Traditional approaches of DSP isolation involved cumbersome and labor-intensive purification strategies either from low-yield native sources such as the rodent brain, or from heterologous yeast, insect, and mammalian cell overexpression systems given to spurious post-translational modifications (and resultant protein heterogeneity). Largely overlooked in the past, the relatively simple expression and purification of a variety of mammalian soluble DSPs under stringently controlled conditions in *Escherichia coli* is now commonplace. However, caveats persist, especially for DSPs such as Mfn and OPA1 (long form) anchored in membranes via transmembrane segments that require more specialized expression systems. Furthermore, most full-length DSPs remain recalcitrant to high-resolution structural characterization largely due to the presence of long, intrinsically disordered regions (IDRs) interposed with folded domains. This is especially true for the mitochondrial dynamins Drp1 (Dnm1), Mfn (Fzo1), and OPA1 (Mgm1), as well as for yeast Vps1. This structural biology bottleneck has been somewhat circumvented by the isolation and crystallization of orthologs from the thermophilic fungus *Chaetomium thermophilum*, which contain more stably folded segments in place of the IDRs. However, whether these structures are truly representative of their foils in non-thermophilic organisms remains an open question.

The first two parts of this protocol series are devoted to methods describing the heterologous expression, purification, and the initial biochemical characterization of the better characterized DSPs from yeast (fungi) and mammals. The third and final part relates to previously undescribed analytical techniques and methodologies geared toward the biophysical and cellular characterization of DSPs in membrane remodeling, fission and fusion.

Cleveland, OH, USA *Rajesh Ramachandran*

Contents

Contributors

FRANCES JOAN D. ALVAREZ • *Department of Structural Biology, University of Pittsburgh School of Medicine, Pittsburgh, PA, USA; Division of Structural Biology, Wellcome Trust Centre for Human Genetics, University of Oxford, Oxford, UK*

ARUN ANANTHARAM • *Department of Pharmacology, University of Michigan, Ann Arbor, MI, USA*

TADATO BAN • *Department of Protein Biochemistry, Institute of Life Science, Kurume University, Fukuoka, Japan*

PAVEL V. BASHKIROV • *Federal Research and Clinical Center of Physical-Chemical Medicine, Moscow, Russia; A.N. Frumkin Institute of Physical Chemistry and Electrochemistry, Russian Academy of Sciences, Moscow, Russia*

BRIANNA L. BAUER • *Department of Pharmacology, Center for Mitochondrial Diseases and Cleveland Center for Membrane and Structural Biology, Case Western Reserve University School of Medicine, Cleveland, OH, USA*

NIKHIL BHARAMBE • *Department of Physiology and Biophysics, Case Western Reserve University School of Medicine, Cleveland, OH, USA*

KSENIA V. CHEKASHKINA • *Federal Research and Clinical Center of Physical-Chemical Medicine, Moscow, Russia; A.N. Frumkin Institute of Physical Chemistry and Electrochemistry, Russian Academy of Sciences, Moscow, Russia*

RYAN W. CLINTON • *Department of Pharmacology, Center for Mitochondrial Diseases and Cleveland Center for Membrane and Structural Biology, Case Western Reserve University School of Medicine, Cleveland, OH, USA*

MARIJN G. J. FORD • *Department of Cell Biology, University of Pittsburgh School of Medicine, Pittsburgh, PA, USA*

VADIM A. FROLOV • *Department of Biochemistry and Molecular Biology, Biophysics Institute (CSIC, UPV/EHU), University of the Basque Country, Leioa, Spain; IKERBASQUE, Basque Foundation for Science, Bilbao, Spain*

CHRISTIAN HERRMANN • *Physical Chemistry I, Faculty of Chemistry and Biochemistry, Ruhr University Bochum, Bochum, Germany*

R. BLAKE HILL • *Department of Biochemistry, Medical College of Wisconsin, Milwaukee, WI, USA*

SUZANNE HOPPINS • *Department of Biochemistry, University of Washington, Seattle, WA, USA*

DI HU • *Departments of Physiology and Biophysics, Case Western Reserve University School of Medicine, Cleveland, OH, USA*

NAOTADA ISHIHARA • *Department of Protein Biochemistry, Institute of Life Science, Kurume University, Fukuoka, Japan; Department of Biological Science, Graduate School of Science, Osaka University, Osaka, Japan*

CAROLYN M. KELLY • *Department of Molecular Medicine, College of Veterinary Medicine, Cornell University, Ithaca, NY, USA*

NOLAN W. KENNEDY • *Department of Biochemistry, Medical College of Wisconsin, Milwaukee, WI, USA*

JESSICA LAIMAN • *Institute of Molecular Medicine, College of Medicine, National Taiwan University, Taipei, Taiwan*

Ya-Wen Liu • *Institute of Molecular Medicine, College of Medicine, National Taiwan University, Taipei, Taiwan; Center of Precision Medicine, National Taiwan University, Taipei, Taiwan*

Patrick J. Macdonald • *Department of Physiology and Biophysics, Case Western Reserve University School of Medicine, Cleveland, OH, USA*

Jason A. Mears • *Department of Pharmacology, Center for Mitochondrial Diseases and Cleveland Center for Membrane and Structural Biology, Case Western Reserve University School of Medicine, Cleveland, OH, USA*

Pooja Madan Mohan • *Department of Biochemistry, Case Western Reserve University School of Medicine, Cleveland, OH, USA*

Felipe Montecinos-Franjola • *Department of Physiology and Biophysics, Case Western Reserve University School of Medicine, Cleveland, OH, USA*

John P. O'Donnell • *Department of Molecular Medicine, College of Veterinary Medicine, Cornell University, Ithaca, NY, USA; Cell Biology Division, MRC Laboratory of Molecular Biology, Cambridge, UK*

Lora K. Picton • *Department of Biochemistry, Medical College of Wisconsin, Milwaukee, WI, USA*

Emily R. Prantzalos • *Department of Pharmacology, University of Michigan, Ann Arbor, MI, USA*

Xin Qi • *Departments of Physiology and Biophysics, Case Western Reserve University School of Medicine, Cleveland, OH, USA; Center for Mitochondrial Disease, Case Western Reserve University School of Medicine, Cleveland, OH, USA*

Rajesh Ramachandran • *Department of Physiology and Biophysics, Case Western Reserve University School of Medicine, Cleveland, OH, USA; Cleveland Center for Membrane and Structural Biology, Case Western Reserve University School of Medicine, Cleveland, OH, USA*

Nyssa Becker Samanas • *Department of Biochemistry, University of Washington, Seattle, WA, USA*

Anna V. Shnyrova • *Department of Biochemistry and Molecular Biology, Biophysics Institute (CSIC, UPV/EHU), University of the Basque Country, Leioa, Spain*

Linda Sistemich • *Physical Chemistry I, Faculty of Chemistry and Biochemistry, Ruhr University Bochum, Bochum, Germany*

Katherine A. Smith • *Department of Pharmacology, University of Michigan, Ann Arbor, MI, USA*

Holger Sondermann • *Department of Molecular Medicine, College of Veterinary Medicine, Cornell University, Ithaca, NY, USA*

Natalia Stepanyants • *Department of Physiology and Biophysics, Case Western Reserve University School of Medicine, Cleveland, OH, USA*

Natalia V. Varlakhanova • *Department of Cell Biology, University of Pittsburgh School of Medicine, Pittsburgh, PA, USA*

Peijun Zhang • *Department of Structural Biology, University of Pittsburgh School of Medicine, Pittsburgh, PA, USA; Division of Structural Biology, Wellcome Trust Centre for Human Genetics, University of Oxford, Oxford, UK; Electron Bio-Imaging Centre, Diamond Light Source, Harwell Science and Innovation Campus, Didcot, UK*

Part I

Yeast DSPs: Isolation and Biochemical Characterization

Chapter 1

Isolation and Analysis of Mitochondrial Fission Enzyme DNM1 from *Saccharomyces cerevisiae*

Nolan W. Kennedy, Lora K. Picton, and R. Blake Hill

Abstract

Mitochondrial fission, an essential process for mitochondrial and cellular homeostasis, is accomplished by evolutionarily conserved members of the dynamin superfamily of large GTPases. These enzymes couple the hydrolysis of guanosine triphosphate to the mechanical work of membrane remodeling that ultimately leads to membrane scission. The importance of mitochondrial dynamins is exemplified by mutations in the human family member that causes neonatal lethality. In this chapter, we describe the subcloning, purification, and preliminary characterization of the budding yeast mitochondrial dynamin, DNM1, from *Saccharomyces cerevisiae*, which is the first mitochondrial dynamin isolated from native sources. The yeast-purified enzyme exhibits assembly-stimulated hydrolysis of GTP similar to other fission dynamins, but differs from the enzyme isolated from non-native sources.

Key words Mitochondrial dynamics, Mitochondrial fission, Dynamin, Drp1, Membrane scission

1 Introduction

Proteins in the dynamin superfamily are large GTPases thought to be mechanoenzymes that harness the energy of GTP hydrolysis to remodel intracellular membranes [1–9]. The dynamins can be classified as either fission or fusion enzymes [10], and studies with recombinant proteins have revealed that a common feature of the fission dynamins is self-assembly, which stimulates GTP hydrolysis [11–23]. Mutations that impair self-assembly and hydrolysis in the human fission dynamins cause severe phenotypic defects, such as, neurological abnormalities and neonatal lethality [2, 24–26]. For example, in the human mitochondrial fission dynamin-1-like protein (DNM1L) (also known as DRP1, DLP1), a spontaneous heterozygous mutation in DNM1L causes neonatal lethality [27] and was subsequently shown to impair protein localization, assembly, and hydrolysis [28]. In this latter study, as in many others, the

Nolan W. Kennedy and Lora K. Picton contributed equally to this work.

Rajesh Ramachandran (ed.), *Dynamin Superfamily GTPases: Methods and Protocols*, Methods in Molecular Biology, vol. 2159,
https://doi.org/10.1007/978-1-0716-0676-6_1, © Springer Science+Business Media, LLC, part of Springer Nature 2020

mitochondrial fission dynamin was heterologously expressed. This has led to detailed mechanistic and structural insights into this important enzyme superfamily [13, 18, 29–31]. In this chapter, we describe the cloning, purification, and preliminary characterization of the mitochondrial dynamin DNM1 from *Saccharomyces cerevisiae* that is the first description of this family member not heterologously expressed. GTP hydrolysis of DNM1 purified from native *S. cerevisiae* compared to that purified from *Escherichia coli* suggests differences in their activities that support isolation from native sources may be important.

2 Materials

2.1 Yeast Strains and Plasmids

1. A modified pEG(KT) plasmid for protein overexpression in *Saccharomyces cerevisiae* [32].
2. *Saccharomyces cerevisiae* strain SEY6210 (*MAT*α, *leu2-3, 112, ura3-52, his3-Δ200, trp1-Δ901, suc2-Δ9, lys2-801; GAL).*
3. A protease-deficient strain of *S. cerevisiae*, DDY1810 (MATa, *leu2Δ, trp1Δ, ura3-52, prb1-1122, pep4-3, pre1-451*) (Shang 2003).

2.2 Chemicals and Reagents

1. Drop-out Mix Synthetic Medium Minus Uracil w/o Yeast Nitrogen Base (US Biologicals, Cat #D9535).
2. Yeast Nitrogen Base w/o AA, Carbohydrate & w/o AS (YNB) (US Biologicals, Cat #Y2030).
3. Carbohydrates: dextrose, galactose, raffinose, maltose.
4. Tryptophan.
5. Bacto™ Peptone.
6. Yeast extract.
7. 4-(2-Aminoethyl) benzenesulfonyl fluoride hydrochloride (AEBSF).
8. Phenylmethylsulfonyl fluoride (PMSF).
9. NaCl.
10. KCl.
11. 4-(2-Hydroxyethyl)-1-piperazineethanesulfonic acid (HEPES).
12. Piperazine-N,N′-bis(2-ethanesulfonic acid) (PIPES).
13. Phosphoenolpyruvate (PEP).
14. NaH_2PO_4.
15. $MgCl_2$.
16. Dithiothreitol (DTT).
17. Dry ice.

18. Ethanol.
19. cOmplete, ULTRA, Mini, EDTA-free protease inhibitors (EASYpack from Roche, Cat. No. 05 892 791 001).
20. Enzymes: DNase (Sigma-Aldrich), Tobacco Etch Virus (TEV) protease (in-house, *see* **Note 1**), pyruvate kinase (Sigma-Aldrich), lactate dehydrogenase (Sigma-Aldrich).
21. Nucleotides: guanosine 5′-triphosphate (GTP), guanosine 5′-[γ-thio]triphosphate (GTP-γ-S) (Sigma-Aldrich).
22. Mdivi-1 ≥98% (HPLC) (Sigma-Aldrich, M0199-25MG).
23. Dynasore hydrate ≥98% (HPLC) (Sigma-Aldrich, D7693-25MG).
24. Bottle-top filter units with 0.2 μm cutoff, 500 mL, for filtering buffer and solutions.
25. Acrodisc® 25 mm syringe filters, 0.45 μm, low-protein binding.
26. 50 kDa MWCO dialysis tubing (Spectra/Por®).

2.3 Buffers and Media

All buffers are made at 25 °C using analytical grade reagents or better with distilled and deionized water (>18 MΩ-cm). For protein isolation and purification, all buffers are stored and used at 4 °C. All buffers should be filtered and degassed.

1. Complete synthetic medium lacking uracil with dextrose (CSM +D–U):
   - (a) 2 g Drop-out Mix Synthetic Medium.
   - (b) g Yeast Nitrogen Base (w/o AA, carbohydrate, and AS).
   - (c) 20 g Glucose (dextrose).
   - (d) 5 g Ammonium sulfate.
   - (e) pH 6.0.
   - (f) Bring to ~900 mL of ddH_2O, stir until dissolved, adjust pH to ~6, and bring to 1 L total volume.
   - (g) Sterile filter (0.2 μm).
2. Complete synthetic medium lacking uracil and leucine (CSM +R–UL), CSM with 2% raffinose plus 40 mg/L tryptophan.
3. Induction medium: 100 mL of 20% galactose, 20 g Bacto™ Peptone, and 10 g yeast extract supplemented to each liter of cell growth.
4. Lysis buffer: 50 mM NaH_2PO_4, pH 7.4, 500 mM NaCl, 5 mM $MgCl_2$, 2 mM DTT.
5. Elution buffer: Lysis buffer plus 20 mM maltose, pH 7.4.
6. Reaction buffer: 125 mM HEPES pH7, 125 mM PIPES pH 7.0, 5 mM $MgCl_2$, 37.5 mM KCl, 5 mM PEP, 100 U/mL pyruvate kinase/lactose dehydrogenase, 1.5 mM NADH, and varying NaCl and 5× GTP as indicated.

7. 5× assay buffer: 125 mM HEPES, 125 mM PIPES, 37.5 mM KCl, 25 mM $MgCl_2$, 5 mM PEP, 3 mM NADH, 100 U/mL pyruvate kinase/lactose dehydrogenase, (pH 7.0). These are stored as single-use aliquots at −20 °C.
8. 4 M NaCl, stored at room temperature.
9. 100× GTP stocks that are stored at various concentrations (0.1 mM to 100 mM) at −20 °C. This range of substrate concentrations is sufficient for determining enzyme kinetics (*see* **Note 2**).
10. DNM1 stocks flash–frozen in a dry ice and ethanol bath and are stored at −80 °C until use. Stocks are not used for multiple freeze–thaw cycles and are not stored for more than a month before use.

2.4 Chromatography Columns

Following columns are used for protein purification and analysis: Dextrin Sepharose® High Performance column (MBTrap, GE Healthcare) and Superose™ 6, 16/60 column (GE Healthcare) (*see* **Note 3**).

2.5 Instrument

1. New Brunswick Shaker incubator.
2. EmulsiFlex-C3 homogenizer (Avestin).
3. Avanti J-20XP centrifuge with 6-L rotor JLA-8.1000 and rotor JA-25.50 (Beckman Coulter) or equivalent.
4. ÄKTA FPLC chromatography system (GE Healthcare).
5. Molecular Devices FlexStation 3 plate reader.
6. Eppendorf Multichannel Pipettes (P2, P200) and pipette tips.

3 Methods

3.1 Subcloning of DNM1 in S. cerevisiae

DNM1 was expressed as a fusion construct with *E. coli* maltose-binding protein (MBP) separated by a Tobacco Etch Virus (TEV) protease-cleavage site (MBP–TEV–DNM1) [33]. A gene encoding MBP–TEV–DNM1 was subcloned into a modified pEG(KT) backbone by homologous recombination in yeast. The pEG (KT) expression vector contains the glutathione S-transferase (GST) gene in the open reading frame [32], which was removed by digesting the vector with Sac*I* followed by agarose gel purification of the pEG (KT) backbone lacking GST gene. The full-length MBP–TEV–DNM1 open reading frame was PCR amplified using MBP–TEV–DNM1 primers as a template [34] and with homology to pEG(KT) at the 5′ and 3′ ends. The MBP–TEV–DNM1 PCR fragment and Sac*I*-digested pEG(KT) were transformed into SEY6210 yeast using the LiAc method [35]. Circularized plasmids were recovered by plasmid rescue, and successful recombination of

MBP–TEV–DNM1 into the pEG(KT) backbone was verified by sequencing. Protease-deficient DDY1810 cells (MATa, *leu2Δ*, *trp1Δ*, *ura3-52*, *prb1-1122*, *pep4-3*, *pre1-451*) [36] were transformed with the pEG(KT)-MBP–TEV–DNM1 plasmid using the LiAc method.

3.2 Expression of DNM1 in *S. cerevisiae*

1. Inoculate starter cultures of MBP–TEV–DNM1 yeast cells in a 250 mL baffled Erlenmeyer flask containing 50 mL of synthetic complete medium lacking uracil and with 2% glucose for 36–48 h at 30 °C with shaking at 200 rpm until reaching OD_{600} of ~2.
2. Dilute the 25 mL starter culture into 1L of complete synthetic medium lacking uracil and leucine, with 2% raffinose and 2× tryptophan. Cells are grown at 26 °C with shaking at 250 rpm until OD_{600} of ~2 (~24 h).
3. Induce protein expression by addition of 100 mL of 20% w/v galactose. The medium is also supplemented at this time with 10 g yeast extract and 20 g Bacto™ Tryptone. Before induction, freeze a 500 μL aliquot of the growth for subsequent SDS-PAGE analysis.
4. Grow cells overnight (typically 16–18 h) at 26 °C with shaking at 250 rpm.
5. Collect cells by centrifugation at 6000 × *g* for 20 min at 4 °C.
6. Wash cell pellet by resuspending in 50 mL of ice-cold sterile water on ice, transfer to two, tared 50 mL conical tubes, and spin at 2500 × *g* at 4 °C for 10 min. Repeat this once and determine mass of the cell pellet and record cell weight in a laboratory notebook (*see* **Note 4**).
7. Resuspend and pool the cell pellets to a final volume of ~80 mL (or less) in ice-cold, sterile water with 4-(2-aminoethyl) benzenesulfonyl fluoride hydrochloride (0.1 mM final). Dispense as 10 mL aliquots into 50 mL conical tubes, freeze on dry ice or ethanol bath, and store at −80 °C until purification.

3.3 Lysis of *S. cerevisiae* Cells

All steps need to be at 4 °C to ensure retention of DNM1 activity.

1. Thaw four 10 mL aliquots of frozen yeast cells on ice, and pool into a 250 mL Erlenmeyer flask using ice-cold lysis buffer to assist in quantitative transfer. Total volume should be ~60–80 mL.
2. Dissolve one tablet of EDTA-free protease inhibitors using a spatula or glass rod. Add PMSF to a final concentration of 2 mM.
3. Lyse the cells at 4 °C by mechanical disruption by 5–6 passes through an EmulsiFlex-C3 homogenizer (Avestin) at

21,000 psi (*see* **Note 5**). Assess 50–200 μL by microscopy to ensure cell lysis (*see* **Note 6**).

4. Add DNase to 1 μg/mL (from 1000× stock), transfer the cell lysate into precooled centrifuge tubes, and clarify by centrifugation at 25,000 × *g* for 20 min.
5. Determine protein concentration of the lysate by Bradford assay.

3.4 Purification of DNM1

All steps need to be at 4 °C including chromatography to ensure retention of DNM1 activity.

1. Procure appropriate amylose affinity column chromatography. Amylose resin capacity is ~3 mg/mL bed volume (NEB), and for a typical 1 L preparation, a 10 mL column suffices.
2. Pre-equilibrate the amylose column with 5–10 column volumes of lysis buffer (*see* **Note 7**). Record baseline A_{280} if using a chromatography system with a UV detector.
3. Filter supernatant of cell lysate using 0.45 μm low-protein binding syringe filter using 80 mL syringe. This step may require more than one syringe filter.
4. Apply filtrate onto the amylose resin at 1 mL/min.
5. Wash the column with 10 column volumes of lysis buffer or until A_{280} returns to baseline value noted above.
6. Elute with 5–10 column volumes of elution buffer (lysis buffer + 20 mM maltose) or until baseline A_{280} is reached. Collect 1–3 mL fractions.
7. Analyze fractions by 10% SDS-PAGE for estimation of protein purity and UV/Vis for A_{280}:A_{260} ratio. DNM1 typically gives purity at this stage of >85% with little if any nucleotide bound giving an A_{280}:A_{260} ratio <1 (*see* **Note 8**). Pool desired fractions and determine protein concentration.
8. Liberate DNM1 from MBP by transferring pooled MBP–TEV–DNM1 fractions into 50 kDa MWCO dialysis tubing (DNM1 = 85 kD, MBP = 45 kD, and TEV = 15 kD), adding TEV protease at 1:10 molar ratio (TEV:MBP–TEV–DNM1), and incubating in 1–2 L of ice-cold lysis buffer for 12–24 h with gentle stirring at 4 °C to cleave the N-terminal MBP tag (*see* **Note 9**).
9. Concentrate the DNM1 using centrifugal concentrating devices with 50 kDa MWCO (*see* **Note 10**).
10. Isolate the DNM1 by size-exclusion chromatography using a Superose™ 6 column (GE Healthcare), discarding protein found in the void volume. Collect 1–3 mL fractions.

11. Assess DNM1 homogeneity of chromatographic fractions by Coomassie-stained 10% SDS-PAGE gel, which is typically found to be at least 90% homogeneous.
12. Single-use aliquots of the DNM1 are frozen at 2–9 mg/mL using a dry ice or ethanol bath and stored at −80 °C until needed (*see* **Note 11**).

3.5 Measurement of GTP Hydrolysis by DNM1

An enzyme-coupled GTPase assay can be used to measure the rate of GTP hydrolysis by DNM1, as described previously [14, 37]. For this assay, the rate of hydrolysis is determined by monitoring NADH depletion at A_{340} using a plate reader. NADH is depleted as pyruvate is converted to lactate via lactate dehydrogenase (LDH). This reaction is coupled to the hydrolysis of GTP to GDP and phosphate by DNM1 and subsequent conversion of phosphoenolpyruvate (PEP) to pyruvate via pyruvate kinase (PK). This assay provides constant regeneration of GTP, maintaining the desired substrate concentration until depletion of NADH. With a multimodal plate reader, this assay can also be used to measure the degree of light scattering by recording the A_{450} signal, which correlates with DNM1 assembly [13]. A further adaptation of this approach has been used to assess assembly and disassembly of human DNM1L [38].

1. Thaw stocks of DNM1 on ice-water bath mixture (*see* **Note 12**).
2. While enzyme is thawing, mix appropriate reagents into a master mix stock so that the final reaction conditions are as follows: 25 mM HEPES, 25 mM PIPES, 7.5 mM KCl, 5 mM $MgCl_2$, 1 mM PEP, 600 μM NADH, and 20 units/mL PK/LDH. NaCl concentration can be varied from 50 mM to 1000 mM (*see* **Note 13**).
3. Once the enzyme stock is thawed, add it to the master mix to the appropriate concentration (usually 1 μM to 20 μM).
4. Incubate the master mix with enzyme on ice for 15 min (*see* **Note 14**).
5. During the incubation, thaw and aliquot GTP at necessary concentrations into a V-bottom 96-well plate (*see* **Note 15**).
6. After the 15 min incubation, aliquot 148.5 μL of the master mix containing DNM1 into a well of a clear-bottom V-bottom 96-well plate. Repeat as necessary (*see* **Note 16**).
7. To start the reaction, pipette 1.5 μL of the GTP stocks into the appropriate reaction wells using an Eppendorf P2 Multichannel Pipette (*see* **Note 17**).
8. Mix the reactions by pipetting up and down with an Eppendorf P200 Multichannel Pipette (*see* **Note 18**).

9. Place the 96-well plate into the plate reader, and set it to measure the absorbance at 340 nm and 450 nm for 1 h at the shortest-allowable intervals.
10. Once measurements are complete, export the data as a plain text (.txt or .csv) file and open in Microsoft Excel or other data analysis software.
11. Plot the absorbance data at 340 nm as a line graph, and determine the rate of hydrolysis for each reaction (*see* **Note 19**).
12. Plot the absorbance data at 450 nm as a line graph, and determine the rate of assembly for each reaction. This is done by fitting the linear range of the data to a straight line over the largest possible window.

3.6 Preliminary Results

Previous studies on DNM1 used purified enzyme from either bacteria [34] or insect cells [13, 14]. In this study, we overexpressed DNM1 in native *Saccharomyces cerevisiae* as an N-terminal fusion construct with a TEV protease-cleavable MBP in protease-deficient strain for purification. Side-by-side GTP hydrolysis experiments confirmed that DNM1 purified from bacteria differed in enzyme properties from yeast-purified enzyme (Fig. 1). Substrate kinetics measurements at various NaCl concentrations indicate that bacterially expressed DNM1 has a higher affinity for substrate than the yeast-expressed enzyme ($K_{0.5}$ of 110 ± 8 μM for bacterial versus 685 ± 47 μM for yeast expressed), but a lower V_{max} (4.0 ± 0.1 nmol/min for bacterial versus 5.2 ± 0.1 nmol/min for yeast expressed). The yeast-expressed enzyme also appears to have weaker substrate affinity and activity than DNM1 derived from insect cells [14]. The basis for these differences might derive from differences in posttranslational modifications, which are known to affect activity of other dynamins [39, 40].

DNM1 isolated from yeast also shows the similar salt-dependent assembly compared to other dynamins. Dynamin-related proteins are known for increasing GTP hydrolysis activity as a function of assembly, which has been stimulated by increasing concentrations of protein or decreasing salt concentrations. Increasing NaCl concentration decreased GTP hydrolysis of DNM1 (Fig. 2). At the highest salt concentration (1 M), the activity was 2.7 nmol/min and increased to 16 nmol/min at the lowest NaCl concentration tested (57 mM). These experiments indicate that DNM1 isolated from yeast is assembly competent with characteristics similar to other dynamins [14, 23, 41, 42]. Finally, the yeast-derived enzyme is also inhibited by the pharmacological inhibitors, Dynasore hydrate (not shown), and Mdivi-1 [43].

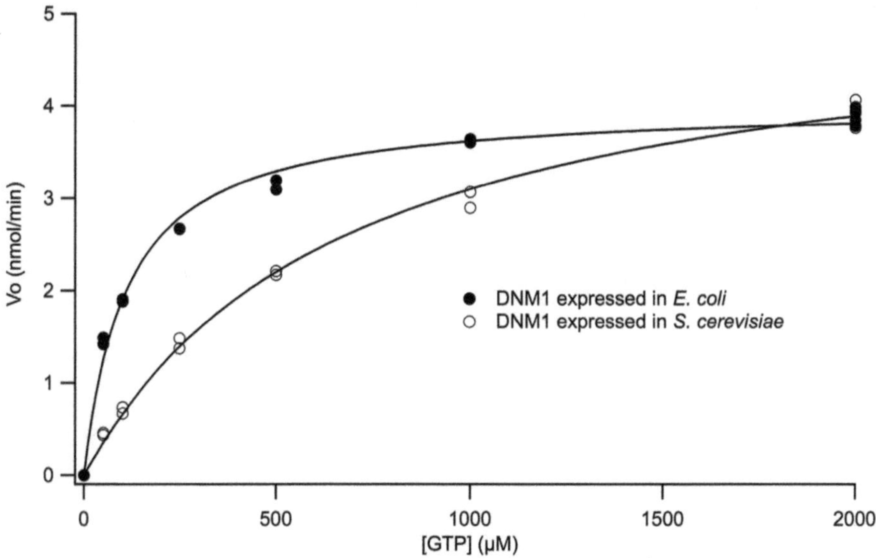

Fig. 1 GTP hydrolysis by DNM1 isolated from *E. coli* (filled circles) and *S. cerevisiae* (open circles). 2 μM DNM1 was incubated with the indicated concentrations of GTP, and hydrolysis measured via NADH depletion using a coupled, regenerative assay that allows for continuous supply of substrate (*see* text). The linear region of the NADH depletion curve was fit to a straight line and converted to initial velocity, V_o, and plotted versus GTP. The data were fit to the generalized Michaelis–Menten equation: $V_o = (V_{max} * [S])/(K_{0.5} + [S])$. For the bacterially expressed DNM1, $K_{0.5} = 110 \pm 8$ μM and $V_{max} = 4.0 \pm 0.1$ nmol/min. For the yeast-expressed DNM1, $K_{M0.5} = 685 \pm 47$ μM and $V_{max} = 5.2 \pm 0.1$ nmol/min. [NaCl] = 233 mM

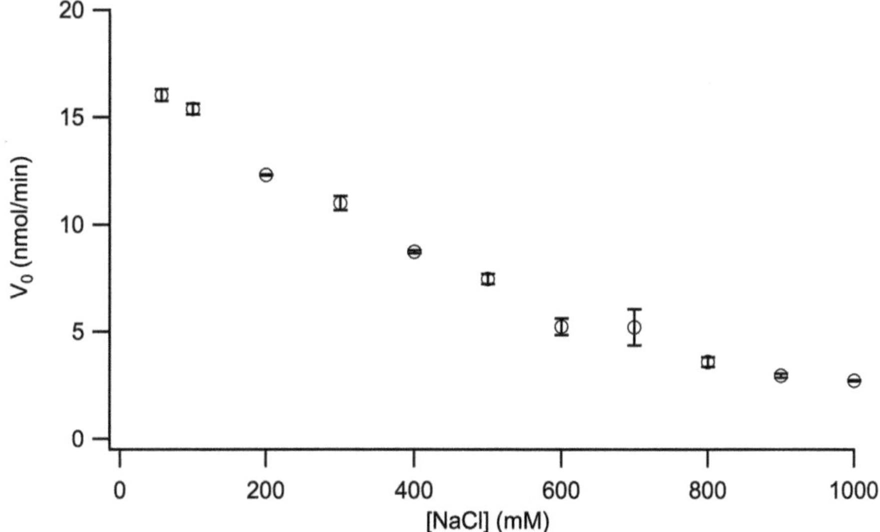

Fig. 2 NaCl dependence on GTP hydrolysis by 10 μM DNM1 isolated from yeast. Dnm1 activity is dependent on NaCl concentration. Initial velocity $V_o \pm$ SE ($n = 3$) is plotted versus [NaCl]

4 Notes

1. We express and purify the catalytic domain variant (S219V) from the Tobacco Etch Virus protease in *E. coli* as developed for protein expression by Waugh and coworkers [32]. This variant resists autoinactivation and is approximately twofold more active than wild type.
2. We typically make up stock solutions fresh or use within 1 month.
3. We have also used amylose resin from NEB with similar success.
4. Greater than 50% increases or decreases in cell pellet mass can signal problems with the growth that can be diagnosed by SDS-PAGE of the whole cell lysate samples taken at induction and harvest.
5. This is the maximum pressure observed during the power stroke.
6. We have not extensively investigated other common methods for yeast cell lysis that include french press, glass beads, and planetary ball mill grinders, with the latter being popularized by Dr. David Rout's laboratory for yielding intact protein complexes.
7. Ensure that the resin is cleaned before use, especially in multi-use laboratories. For amylose resins, we wash with 5 column volumes of 20 mM maltose, 5 column volumes of water, 3 column volumes of 0.1% SDS, 10 column volumes of water, 5 column volumes of 20% EtOH.
8. DNM1 is typically purified without nucleotide bound unlike small GTPases. If nucleotide is present, procedures can generate nucleotide-free GTPase [44].
9. A 1 or 2 L graduated cylinder works well for this purpose.
10. DNM1 self-assembly can occlude the pores in centrifugal devices. This can be minimized by using 500 mM NaCl and <10 mg/mL protein concentrations that favor disassembly. The next purification step is size-exclusion chromatography, in which a minimal volume is desirable, so we typically concentrate down to 1 mL or 2 mL max.
11. The final concentration depends on the NaCl concentration and other considerations. Some mutants affect assembly of DNM1 and can be concentrated to >20 mg/mL. For DNM1 concentration determination, the protein is unfolded with 6 M GdnHCl, and A_{280} measured. The protein concentration is calculated from Beer's law using the theoretical extinction coefficient (38,280 M^{-1} cm^{-1}).

12. We find that rapid thawing at room temperature leads to enzyme aggregation. Once the enzyme is thawed, it degrades over time and can aggregate within 6 h. If aggregation is apparent, we discard that stock.
13. Consistent with other dynamins [11, 42], we find that low NaCl (50 mM) leads to the highest levels of hydrolysis or assembly, while high NaCl (1000 mM) leads to the lowest levels of hydrolysis or assembly (Fig. 2).
14. The rationale of the 15-min incubation is that some variation exists in how long it takes to set up the GTP stocks or salt titration plate, depending on what conditions are being tested. Regardless, this never takes more than 15 min. GTP-γ-S is used as a control for contaminating NTPases that might hydrolyze GTP.
15. V-bottom 96-well plate plates are used for rapid transfer of the substrate into the reaction plate to start the reaction. Use of the 96-well plate allows transfer with an Eppendorf Multichannel Pipette. We find that a minimum of 5 μL is necessary to ensure accurate pipetting from the V-bottom plate into the reaction plate.
16. Pipette carefully to avoid bubble formation, which can interfere with absorbance measurements. If present, use a narrow gauge needle or pipette tip to disrupt them.
17. Using multichannel pipettes is essential but can result in inaccuracies of one or more channels, which needs to be visually checked before dispensing. When this occurs, the solution is placed back into the reservoir, and all tips discarded. Inaccuracies can be minimized by ensuring a good seal between the pipette tip and pipettor.
18. Make sure to do this step as quickly as possible, as reactions under high enzyme, assembly, and/or substrate conditions can rapidly deplete NADH.
19. This is done by fitting the linear range of the data to a straight line over the largest possible window. Convert slope (ΔA_{340}/s) to activity in nmol/min with the following equation: activity = (ΔA_{340}/s $*$60 s/min) $*$ (0.001 L) $*$ (1^9 nmol/mol)/ (6220 M^{-1} cm^{-1})/(0.41 cm). A buffer control for each GTP concentration is also collected, and a background hydrolysis rate calculated and subtracted from the activity.

Acknowledgement

The reagents pEG(KT), SEY6210, and DDY1810 were kind gifts from B. Wendland. This work was supported by the National Institutes of Health grant R01GM067180.

References

1. Ramachandran R, Schmid SL (2018) The dynamin superfamily. Curr Biol 28: R411–R416. https://doi.org/10.1016/j.cub.2017.12.013
2. Faelber K, Gao S, Held M et al (2013) Oligomerization of dynamin superfamily proteins in health and disease. Prog Mol Biol Transl Sci 117:411–443. https://doi.org/10.1016/B978-0-12-386931-9.00015-5
3. Praefcke GJ, McMahon HT (2004) The dynamin superfamily: universal membrane tubulation and fission molecules? Nat Rev Mol Cell Biol 5:133–147
4. Ford M, Nunnari J, Jenni S (2012) An integrated structural analysis of dynamin assembly. Microsc Microanal 18:48–49. https://doi.org/10.1017/S1431927612002097
5. Bui HT, Shaw JM (2013) Dynamin assembly strategies and adaptor proteins in mitochondrial fission. Curr Biol 23:R891–R899. https://doi.org/10.1016/j.cub.2013.08.040
6. Danino D, Hinshaw JE (2001) Dynamin family of mechanoenzymes. Curr Opin Cell Biol 13:454–460
7. Sesaki H, Jensen RE (1999) Division versus fusion: Dnm1p and Fzo1p antagonistically regulate mitochondrial shape. J Cell Biol 147 (4):699–706
8. Smirnova E, Shurland DL, Ryazantsev SN, van der Bliek AM (1998) A human dynamin-related protein controls the distribution of mitochondria. J Cell Biol 143(2):351–358
9. van der Bliek AM (1999) Functional diversity in the dynamin family. Trends Cell Biol 9 (3):96–102
10. Lee MW, Lee EY, Lai GH et al (2017) Molecular motor Dnm1 synergistically induces membrane curvature to facilitate mitochondrial fission. ACS Cent Sci 3:1156–1167. https://doi.org/10.1021/acscentsci.7b00338
11. Leonard M, Doo Song B, Ramachandran R, Schmid SL (2005) Robust colorimetric assays for Dynamin's basal and stimulated GTPase activities. In: Methods in enzymology. Academic Press, New York, pp 490–503
12. Antonny B, Burd C, De Camilli P et al (2016) Membrane fission by dynamin: what we know and what we need to know. EMBO J 35:2270–2284. https://doi.org/10.15252/embj.201694613
13. Mears JA, Lackner LL, Fang S et al (2011) Conformational changes in Dnm1 support a contractile mechanism for mitochondrial fission. Nat Struct Mol Biol 18:20–26. https://doi.org/10.1038/nsmb.1949
14. Ingerman E, Perkins EM, Marino M et al (2005) Dnm1 forms spirals that are structurally tailored to fit mitochondria. J Cell Biol 170:1021–1027
15. Chappie JS, Mears JA, Fang S et al (2011) A pseudoatomic model of the dynamin polymer identifies a hydrolysis-dependent powerstroke. Cell 147:209–222. https://doi.org/10.1016/j.cell.2011.09.003
16. Chappie JS, Acharya S, Leonard M et al (2010) G domain dimerization controls dynamin's assembly-stimulated GTPase activity. Nature 465:435–440. https://doi.org/10.1038/nature09032
17. Otsuga D, Keegan BR, Brisch E et al (1998) The dynamin-related GTPase, Dnm1p, controls mitochondrial morphology in yeast. J Cell Biol 143:333–349
18. Reubold TF, Faelber K, Plattner N et al (2015) Crystal structure of the dynamin tetramer. Nature 525:404. https://doi.org/10.1038/nature14880
19. Wenger J, Klinglmayr E, Frohlich C et al (2013) Functional mapping of human dynamin-1-like GTPase domain based on x-ray structure analyses. PLoS One 8:e71835. https://doi.org/10.1371/journal.pone.0071835
20. Francy CA, Frölich C, Daumke O, Mears JA (2014) Examining Drp1 conformational changes and domain interactions in the mitochondrial fission complex using Cryo-Em. Biophys J 106:601a. https://doi.org/10.1016/j.bpj.2013.11.3328
21. Francy CA, Clinton RW, Fröhlich C et al (2017) Cryo-EM studies of Drp1 reveal cardiolipin interactions that activate the helical oligomer. Sci Rep 7:10744. https://doi.org/10.1038/s41598-017-11008-3
22. Stepanyants N, Macdonald PJ, Francy CA et al (2015) Cardiolipin's propensity for phase transition and its reorganization by dynamin-related protein 1 form a basis for mitochondrial membrane fission. Mol Biol Cell 26:3104–3116. https://doi.org/10.1091/mbc.E15-06-0330
23. Francy CA, Alvarez FJD, Zhou L et al (2015) The mechanoenzymatic core of dynamin-related protein 1 comprises the minimal machinery required for membrane constriction. J Biol Chem 290:11692–11703. https://doi.org/10.1074/jbc.M114.610881

24. Kenniston JA, Lemmon MA (2010) Dynamin GTPase regulation is altered by PH domain mutations found in centronuclear myopathy patients. EMBO J 29:3054–3067. https://doi.org/10.1038/emboj.2010.187
25. Bitoun M, Bevilacqua JA, Eymard B et al (2009) A new centronuclear myopathy phenotype due to a novel dynamin 2 mutation. Neurology 72:93–95. https://doi.org/10.1212/01.wnl.0000338624.25852.12
26. von Spiczak S, Helbig KL, Shinde DN et al (2017) DNM1 encephalopathy: a new disease of vesicle fission. Neurology 89:385–394. https://doi.org/10.1212/WNL.0000000000004152
27. Waterham HR, Koster J, van Roermund CW et al (2007) A lethal defect of mitochondrial and peroxisomal fission. N Engl J Med 356:1736–1741
28. Chang C-R, Manlandro CM, Arnoult D et al (2010) A lethal de novo mutation in the middle domain of the dynamin-related GTPase Drp1 impairs higher order assembly and mitochondrial division. J Biol Chem 285:32494–32503. https://doi.org/10.1074/jbc.M110.142430
29. Faelber K, Posor Y, Gao S et al (2011) Crystal structure of nucleotide-free dynamin. Nature 477:556–560. https://doi.org/10.1038/nature10369
30. Frohlich C, Grabiger S, Schwefel D et al (2013) Structural insights into oligomerization and mitochondrial remodelling of dynamin 1-like protein. EMBO J 32:1280–1292. https://doi.org/10.1038/emboj.2013.74
31. Faelber K, Held M, Gao S et al (2012) Structural insights into dynamin-mediated membrane fission. Structure 20:1621–1628. https://doi.org/10.1016/j.str.2012.08.028
32. Mitchell DA, Marshall TK, Deschenes RJ (1993) Vectors for the inducible overexpression of glutathione S-transferase fusion proteins in yeast. Yeast 9:715–722. https://doi.org/10.1002/yea.320090705
33. Kapust RB, Tözsér J, Fox JD et al (2001) Tobacco etch virus protease: mechanism of autolysis and rational design of stable mutants with wild-type catalytic proficiency. Protein Eng 14:993–1000
34. Wells RC, Picton LK, Williams SC et al (2007) Direct binding of the dynamin-like GTPase, Dnm1, to mitochondrial dynamics protein Fis1 is negatively regulated by the Fis1 N-terminal arm. J Biol Chem 282:33769–33775. https://doi.org/10.1074/jbc.M700807200
35. Schiestl RH, Gietz RD (1989) High efficiency transformation of intact yeast cells using single stranded nucleic acids as a carrier. Curr Genet 16:339–346
36. Koppenol-Raab M, Harwig MC, Posey AE (2016) A targeted mutation identified through pKa measurements indicates a post-recruitment role for Fis1 in yeast mitochondrial fission. J Biol Chem 291(39):20329–20344
37. Ingerman E, Nunnari J (2005) A continuous, regenerative coupled GTPase assay for dynamin-related proteins. In: Methods in enzymology. Academic Press, New York, pp 611–619
38. Cahill TJ, Leo V, Kelly M et al (2016) Resistance of dynamin-related protein 1 oligomers to disassembly impairs mitophagy, resulting in myocardial inflammation and heart failure. J Biol Chem 291:25762. https://doi.org/10.1074/jbc.A115.665695
39. Chang CR, Blackstone C (2007) Cyclic AMP-dependent protein kinase phosphorylation of Drp1 regulates its GTPase activity and mitochondrial morphology. J Biol Chem 282:21583–21587. https://doi.org/10.1074/jbc.C700083200
40. Xie W, Adayev T, Zhu H et al (2012) Activity-dependent phosphorylation of dynamin 1 at serine 857. Biochemistry 51:6786–6796. https://doi.org/10.1021/bi2017798
41. Warnock DE, Hinshaw JE, Schmid SL (1996) Dynamin self-assembly stimulates its GTPase activity. J Biol Chem 271:22310–22314. https://doi.org/10.1074/jbc.271.37.22310
42. Koirala S, Guo Q, Kalia R et al (2013) Interchangeable adaptors regulate mitochondrial dynamin assembly for membrane scission. Proc Natl Acad Sci U S A 110:E1342–E1351. https://doi.org/10.1073/pnas.1300855110
43. Bordt EA, Clerc P, Roelofs BA et al (2017) The putative Drp1 inhibitor mdivi-1 is a reversible mitochondrial complex I inhibitor that modulates reactive oxygen species. Dev Cell 40:583–594.e6. https://doi.org/10.1016/j.devcel.2017.02.020
44. Jaiswal M, Dubey BN, Koessmeier KT et al (2012) Biochemical assays to characterize Rho GTPases. In: Rivero F (ed) Rho GTPases: methods and protocols. Springer, New York, pp 37–58

Chapter 2

Purification of the Dynamin-Related Protein Vps1 Using Mammalian and Bacterial Expression Systems

Natalia V. Varlakhanova and Marijn G. J. Ford

Abstract

The dynamin-related proteins (DRPs) are self-assembling membrane remodeling machines that are indispensable for fundamental cellular trafficking and homeostatic processes. We describe in this chapter protocols developed in our laboratory for purification of full-length and minimal constructs of *Chaetomium thermophilum* Vps1, the model fungal DRP, using mammalian and *Escherichia coli* expression systems.

Key words Vps1, DRP, Dynamin-Related Protein, Purification, Expi293F™, BL21 (DE3)

1 Introduction

The dynamin-related proteins (DRPs) are a family of large GTPases that perform membrane remodeling functions throughout the cell and are essential for many fundamental biological processes that depend on membrane remodeling [1]. These include endocytosis, mitochondrial fusion and fission, trafficking, among others [2–4]. The importance of the DRPs in higher eukaryotes is underscored by their association with several severe human diseases, including neurodegenerative diseases and genetic disorders, such as Charcot–Marie–Tooth type 2A [5].

All DRPs share a conserved domain organization and are believed to remodel membranes by self-assembly coupled with GTP hydrolysis-dependent conformational changes [1, 4]. All include a large GTPase domain, a three-helix bundle known as the bundle signaling element (BSE) that acts as an intramolecular signaling and structural element, and a Stalk that is responsible for DRP self-assembly. Dynamin, uniquely, also contains a PH domain at the base of its Stalk, whereas all others have a variable region of low complexity, termed Insert B or the Variable Domain.

With some notable exceptions, DRPs have been challenging to purify for biochemical and structural analyses. This is largely due to

Rajesh Ramachandran (ed.), *Dynamin Superfamily GTPases: Methods and Protocols*, Methods in Molecular Biology, vol. 2159, https://doi.org/10.1007/978-1-0716-0676-6_2,

their biochemical properties: multi-domain composition, propensities to self-assemble, various interposed regions of low complexity, and their coupling of sometimes rapid GTP hydrolysis with conformational changes.

Vps1 is a conserved and versatile fungal DRP that has been implicated in several membrane remodeling processes. This includes both vacuolar fusion and fission, endosomal and peroxisomal dynamics, and endocytosis and retrograde trafficking [6–12]. As Vps1 shares significant sequence and structural homology with other DRPs, and as it has an Insert B sequence, Vps1 is an attractive model to study DRP biology [6]. Insight into the biochemical and structural determinants of Vps1 self-assembly and analysis of differences compared to other DRPs will enhance understanding of how DRPs are optimized for their specific cellular targets and functions.

In this chapter, we describe protocols developed in our laboratory that can be used to purify full-length Vps1 as well as minimal GTPase-BSE constructs [6]. To purify full-length Vps1, we overexpress it in Expi293F™ Cells, followed by an affinity purification using a maltose-binding protein (MBP) tag. The purified protein self-assembles in vitro and can be used for biochemical and structural analyses. The minimal GTPase-BSE is expressed and purified from *Escherichia coli* and has been used for biochemical and structural analyses.

2 Materials

2.1 Transfection of Expi293F™ Cells

1. Sterile Nalgene™ Single-Use Erlenmeyer Flasks with baffled bottom, either 125 ml or 250 ml.
2. Expi293™ Expression Medium (Gibco™).
3. Opti-MEM I Reduced-Serum Medium (Gibco™).
4. ExpiFectamine™ 293 Transfection Kit (Gibco™).
5. Phosphate-Buffered Saline, 10× Solution (Thermo Fisher Scientific).
6. Expi293F™ Cells (Life Technologies).

2.2 Purification of Vps1 from Expi293F™ Cells

1. Econo-Pac® Chromatography Columns (1–20 ml) (Bio-Rad).
2. Tight-fitting Dounce homogenizer (20 ml).
3. Amylose resin (NEB).
4. Amicon® Ultra-4 Centrifugal Filter Unit 10,000 MWCO (EMD Millipore).
5. Buffer A: 50 mM HEPES pH 8.0, 500 mM NaCl, 5% glycerol, 1% Tween®, 0.3% NP-40, 50 μg/ml DNase I, 5 mM $MgCl_2$,

and 10 mM β-mercaptoethanol, 1× HALT Protease Inhibitors Cocktail (Thermo Fisher Scientific).

6. Incomplete Buffer A: 50 mM HEPES pH 8.0, 500 mM NaCl, and 5% glycerol.
7. Buffer B: 50 mM HEPES pH 8.0, 1 M NaCl, 5% glycerol, and 10 mM β-mercaptoethanol, freshly added.

2.3 Transformation and Growth of E. coli BL21 (DE3)

1. *E. coli* BL21 (DE3) (Novagen).
2. LB media and plates: 10 g Bacto™ Tryptone, 5 g yeast extract, and 10 g NaCl per liter, autoclaved. For plates, add 15 g Bacto™ Agar per liter prior to autoclaving.
3. 2xTY media: 16 g Bacto™ Tryptone, 10 g yeast extract, 5 g NaCl per liter, autoclaved.
4. Isopropyl-β-D-thiogalactopyranoside (Thermo Fisher Scientific): make a stock of 35 mg/ml in water and filter-sterilize.
5. Ampicillin (Thermo Fisher Scientific): make a stock of 100 mg/ml in water and filter-sterilize.

2.4 Purification of Vps1 GTPase-BSE

1. Resuspension buffer: 20 mM Tris Cl pH 8.0, 150 mM NaCl, and 1.93 mM β-mercaptoethanol.
2. Protino® Ni-IDA Resin (Macherey-Nagel™).
3. Protino® Glutathione Agarose 4B (Macherey-Nagel™).
4. PreScission Protease (GE Healthcare Life Sciences™).
5. Amicon® Ultra-15 and Ultra-4 Centrifugal Filter Units of molecular weight cutoff 30,000 and 10,000 Da respectively (EMD Millipore).
6. Slide-A-Lyzer™ Dialysis Cassettes, 30 ml, molecular weight cutoff 10,000 Da (Thermo Fisher Scientific).
7. Q Sepharose® Fast Flow (GE Healthcare).
8. Superdex S200 16/60 gel filtration column (GE Healthcare).

3 Methods

3.1 Transfection of Expi293F™ Cells

1. Grow suspension-adapted Expi293F Cells in Expi293™ Expression Medium, in the absence of antibiotics, to a density of 3.5–4.0 × 10^6 cells/ml (*see* **Note 1**). Grow cells at 37 °C with 8% CO_2, with gentle rotation on an orbital shaker (125 rpm). Cells must have a viability of > 95%. Assess viability by determining exclusion of Trypan Blue (Thermo Fisher Scientific) vital dye. Nonviable cells do not exclude the dye and turn blue. Count cells using a hemocytometer.
2. Dilute a total of 22.5 × 10^7 cells into the fresh prewarmed Expression Medium to a final volume of 76.5 ml. The final cell density is ~2.5 × 10^6 cells/ml. Use a 250 ml Nalgene™ Single-Use Erlenmeyer Flask.

3. While preparing the transfection mixtures, place the cells back onto the orbital shaker in the incubator.
4. Dilute and gently mix 90 μg of high-purity plasmid DNA (*see* **Note 2**) (0.5–5 μg/μl, purified using the GeneJET Plasmid Midi Prep Kit – Thermo Fisher Scientific) into Opti-MEM medium to a total final volume of 4.5 ml. The Opti-MEM must be prewarmed to room temperature.
5. In a separate 15 ml sterile conical tube, dispense 4.26 ml Opti-MEM. Add 240 μl of ExpiFectamine reagent into the Opti-MEM and gently mix. The final volume of the ExpiFectamine mixture is 4.5 ml. Incubate the mixture for 5 min at room temperature.
6. Add the DNA–Opti-MEM mixture (4.5 ml) to the ExpiFectamine–Opti-MEM mixture (4.5 ml) to a total volume of 9 ml. Incubate at room temperature for 25 min.
7. Add this DNA–ExpiFectamine mix to Expi293F™ Cells to a total volume of 85.5 ml.
8. Incubate the cells at 37 °C, with agitation (125 rpm) and with 8% CO_2.
9. After 18 h, add 450 μl of Transfection Enhancer I and 4.5 ml of Transfection Enhancer II to the flask containing the cells.
10. Harvest the cells 24 h after the addition of the enhancers (42 h post-transfection) by centrifugation at 100 × *g* in 50 ml conical tubes for 10 min at 4 °C.
11. Gently wash the cell pellet with ice-cold 1× Phosphate-Buffered Saline (50 ml total). Centrifuge at 100 × *g* for 10 min at 4 °C. Gently decant the supernatant. The cell pellet can be flash-frozen in liquid nitrogen at this stage for storage at −80 °C.

3.2 Purification of Full-length MBP-Tagged Vps1 from Transfected Expi293F™ Cells

1. Thaw out the cell pellet. Gently resuspend the pellet in 20 ml Complete Buffer A. Place the tube on a rocker at 4 °C and incubate for 1 h.
2. Homogenize the lysate using an ice-cold tight-fitting Dounce homogenizer (15 strokes) (*see* **Note 3**).
3. Centrifuge the homogenate at 21,000 × *g* at 4 °C for 30 min in a Hitachi S50A fixed-angle rotor, or equivalent (*see* **Note 4**).
4. Collect the supernatant and mix with 1 ml of a 50% amylose agarose resin (NEB) slurry, prewashed with Complete Buffer A.
5. Incubate the mixture for 1 h at 4 °C on a rocker.
6. Transfer the bead–supernatant mixture to an Econo-Pac® Chromatography Column (Bio-Rad) at 4 °C and let the supernatant flow through by gravity.
7. Wash the beads with 25 ml ice-cold Incomplete Buffer A (no detergents or protease inhibitors).

8. Wash the beads with 12.5 ml ice-cold Buffer B.
9. Wash the beads with 5 ml ice-cold Incomplete Buffer A.
10. Stop the flow by capping the column. Add 2.5 ml of 100 mM maltose in Incomplete Buffer A supplemented with protease inhibitors. Incubate for 15 min at 4 °C without rocking,
11. Collect the flow-through by gravity as Elution Fraction I.
12. Add 2.5 ml of 100 mM maltose in Incomplete Buffer A and collect flow- through as Elution Fraction II.
13. Combine Elution Fractions I and II and add Halt Protease Inhibitor Cocktail to a final concentration of 1×.
14. Assess the quality of the protein by SDS-PAGE and staining with Coomassie Brilliant Blue (Fig. 1).
15. Concentrate the eluted protein using an Amicon® Ultra-4 Centrifugal Filter Unit (10,000 MWCO) (MilliporeSigma), aliquot, and snap-freeze in liquid nitrogen for storage at −80 °C (*see* **Note 5**).

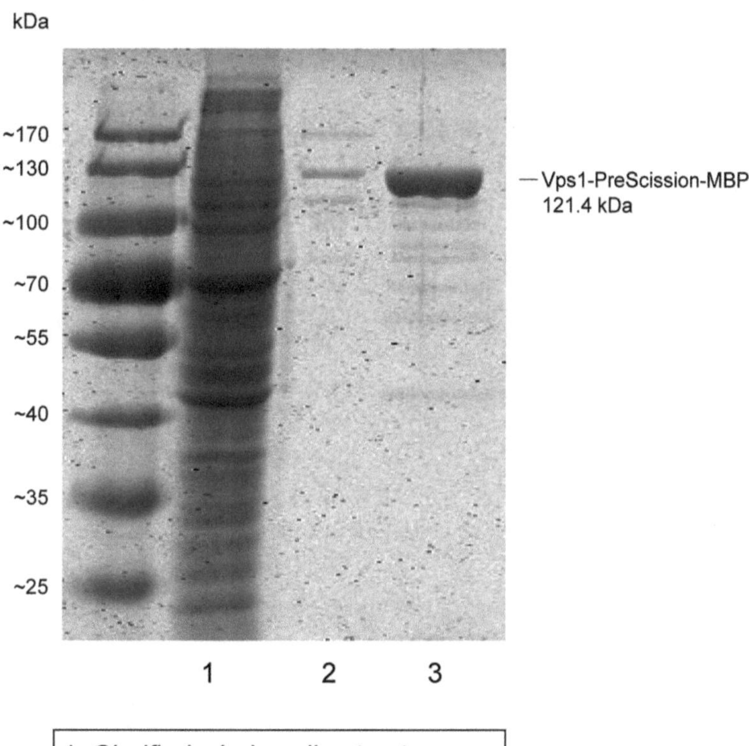

Fig. 1 Purification of full-length MBP-tagged Vps1 from transfected Expi293F™ Cells. A representative SDS-PAGE gel, stained with Coomassie Brilliant Blue, from a purification

3.3 Purification of the Minimal GTPase-BSE Construct of Vps1 from E. coli

1. Transform *Escherichia coli* strain BL21 (DE3) (Novagen) with the GTPase-BSE construct (pET-15b PreScission Vps1 GTPase-BSE) (*see* **Note 6**).
2. Inoculate two to four 50 ml cultures of LB supplemented with 100 μg/ml ampicillin (*see* **Note 7**).
3. Grow the cultures at 37 °C on an orbital shaker (180 rpm) until the culture is saturated (*see* **Note 8**).
4. Transfer each entire 50 ml culture into 1 l 2xTY, in a 2 l Erlenmeyer flask.
5. Grow at 37 °C on an orbital shaker (180 rpm) until the culture reaches an optical density of OD_{600} ~0.6.
6. Induce expression by addition of IPTG to a final concentration of 42.5 μM (*see* **Note 9**).
7. Reduce the ambient temperature of the incubator to 21 °C, and grow the cells overnight (~14 additional h). The final OD_{600} should be ~4.0.
8. Harvest the cells by transferring each liter of culture to a 1 l centrifuge bottle, and centrifuge for 15 min at 3800 × *g* in a Piramoon F9-4x1000y rotor, or equivalent, at 4 °C.
9. Decant the supernatant (*see* **Note 10**), and drain all excess growth medium from the pellet.
10. Resuspend the pellet in ~20 ml Resuspension Buffer per pellet generated from each liter of culture.
11. Snap-freeze in liquid nitrogen and store at −80 °C or proceed to lysis.

3.4 Lysis Using an Avestin Emulsiflex C3 Homogenizer

1. Fully thaw the pellet at 4 °C. Prepare additional ice-cold Resuspension Buffer and ice-cold water.
2. Prechill the sample chamber on ice or by passing ice-cold water through it. Turn on the heat exchanger or place the sample loop in a wet ice-water bath.
3. Switch on the homogenizer and turn on the air supply. Input air pressure needs to be ~100 psi (*see* **Note 11**).
4. Sequentially pass ice-cold water and Resuspension Buffer through the homogenizer with no pressure.
5. Refill the sample chamber with ice-cold Resuspension Buffer. With the machine running, slowly increase the input air pressure using the air pressure regulator until the homogenizing pressure gauge begins pulsing (input pressure ~40 psi).
6. Allow the remaining buffer to flow through the homogenizer.
7. Pour the fully thawed cell pellet into the sample chamber.
8. With the homogenizer running, turn the pressure regulator until it is pulsing at ~15,000 psi (*see* **Note 12**).

9. Collect the lysate in a receptacle chilled on ice.
10. Pass the lysate through the homogenizer an additional 2 times.
11. Flow an additional ~40 ml of Resuspension Buffer through the homogenizer at pressure to collect excess lysate.
12. Clarify the lysate by centrifugation at 142,000 × *g* for 45 min at 4 °C in a Sorvall™ T-647.5 rotor, or similar.

3.5 Purification of Vps1 GTPase-BSE from the Clarified Lysate

1. Incubate the clarified lysate with an appropriate volume of washed Ni-IDA beads (Macherey-Nagel™), prerinsed with Resuspension Buffer (*see* **Note 13**).
2. Incubate the beads with gentle rotation at 4 °C for 1 h in an appropriate number of 50 ml conical tubes.
3. Sediment the beads by centrifugation (1000 × *g* at 4 °C in 5 min) in a refrigerated benchtop centrifuge (*see* **Note 14**).
4. Wash the pellet twice in ice-cold Resuspension Buffer.
5. Pack the beads into a column of appropriate bed volume.
6. Using an FPLC system, wash the beads extensively with Resuspension Buffer using a flow-rate of 2 ml/min, until no more protein washes off the beads.
7. Elute the protein with a shallow gradient from 0 to 250 mM imidazole in Resuspension Buffer.
8. Analyze the peak fractions by running an SDS-polyacrylamide gel followed by staining with Coomassie Brilliant Blue (Fig. 2a).
9. Pool the peak fractions, and add PreScission Protease (GE Healthcare –10 units per 50 ml).
10. Dialyze the pooled peak fractions containing the PreScission Protease into the Resuspension Buffer using Slide-A-Lyzer™ Dialysis Cassettes. Dialyze with gentle stirring at 4 °C into 2 l fresh Resuspension Buffer overnight.
11. Dialyze into 2 fresh Resuspension Buffer for 2 h.
12. Extract samples from the dialysis at various time points for analysis on the progress of cleavage by SDS-polyacrylamide gel electrophoresis (*see* **Note 15**).
13. Once the cleavage is complete, remove the PreScission Protease (which is GST-tagged) by incubation with Protino® Glutathione Agarose 4B (Macherey-Nagel™).
14. Incubate the sample with fresh Ni-IDA beads to remove any remaining non-cleaved sample for 30 min at 4 °C with gentle rotation.
15. Pass the supernatant over a Q Sepharose® anion exchange column (GE Healthcare) in the Resuspension Buffer. The cleaved sample will flow through. Remaining contaminants

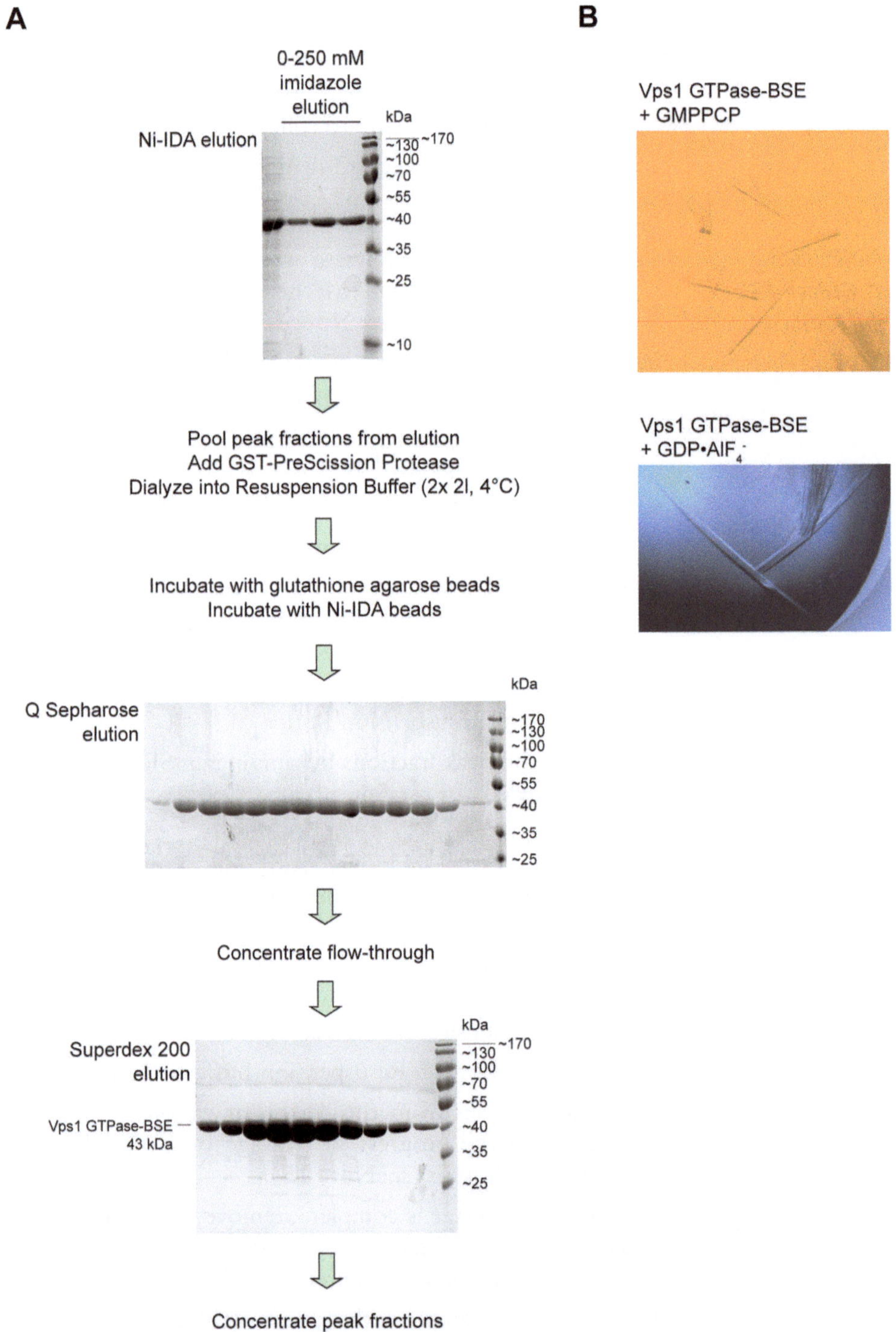

Fig. 2 Purification of Vps1 GTPase-BSE. (**a**) Purification pipeline for Vps1 GTPase-BSE, showing representative SDS-PAGE gels, all stained with Coomassie Brilliant Blue, from the indicated stages of the purification. The gel from the Ni-IDA elution shows every second fraction of the peak as a 1:6 dilution. The gel from the Q Sepharose® shows each fraction from the flow-through as a 1:2 dilution. The gel from the Superdex S200

may be eluted from the column with a salt gradient from 150 to 500 mM NaCl.

16. Analyze the flow-through protein-containing fractions by SDS-polyacrylamide gel electrophoresis followed by Coomassie staining (Fig. 2a).
17. Concentrate the peak fractions by centrifugation at 4500 × *g* at 4 °C using an Amicon® Ultra-15 Centrifugal Filter Unit with a molecular weight cutoff of 30,000 Da.
18. Pass the concentrated protein over a Superdex S200 16/60 size-exclusion column (GE Healthcare) at 1 ml/min, using thoroughly degassed Resuspension Buffer as the mobile phase.
19. Analyze the peak fractions by SDS-polyacrylamide gel electrophoresis followed by Coomassie staining (Fig. 2a).
20. Concentrate the peak fractions as above using an Amicon® Ultra-15 Centrifugal Filter Unit.
21. Aliquot, determine the concentration by measuring the absorption at A_{280}, and snap-freeze in liquid nitrogen for future use (*see* **Note 16**). The protein is appropriate for structural studies (Fig. 2b).

4 Notes

1. Mammalian cells may harbor unknown viruses or pathogens that may pose a significant health risk. Consequently, all manipulation should be conducted in accordance with local requirements and regulations. Appropriately disinfect and dispose of the washed supernatants and solid waste.
2. We use constructs cloned into the pcDNA™3.1 (+) vector (Invitrogen), which drives transcription from the CMV promoter and enhancer. For expression of MBP-tagged Vps1, we typically use a construct encoding Vps1 fused to a PreScission site and the MBP tag at its C-terminus.
3. Avoid bubbles and foaming: proteins denature at air–water interfaces.
4. The pellet on centrifugation is very delicate and gloopy.
5. In general, we obtain ~250 µl of 8–16 µM (1–2 mg/ml) purified MBP-tagged Vps1 per prep.
6. The constructs we use for expression of Vps1 GTPase-BSE introduce a PreScission Protease (HRV 3C) cleavage site

Fig. 2 (continued) column elution shows each fraction of the peak as a 1:2 dilution. (**b**) The purified Vps1 GTPase-BSE can be used for biochemical and structural analysis. Shown are crystals obtained in the presence of Mg^{2+} and the nucleotide analogue GMPPCP (top) and Mg^{2+} and GDP:AlF_4^- (bottom)

between the thrombin site present in pET-15b and the start of the construct coding sequence. A single glycine is used as the linker between the PreScission site and the coding sequence of the construct. This results in an N-terminal overhang of only three amino acids (gly–pro–gly) after cleavage.

7. Use only freshly transformed colonies. Old colonies grow poorly.
8. A 50 ml culture inoculated with a fresh transformant will be saturated after approximately 8 h.
9. Take culture samples immediately prior to induction, after ~2 h, and after overnight induction to aid in monitoring the progress of overexpression and for troubleshooting, if needed.
10. Disinfect the supernatant with 10% v/v bleach, or an equivalent approved disinfectant.
11. If the input air pressure is too low, use an in-line air pressure booster. We use the VBA10A-T02-Z (SMC Pneumatics) that is compatible with the Avestin inlets and outlets.
12. Do not allow the sample chamber to run dry as the introduction of air into the homogenizer will block it. Sample retrieval is possible, but mechanical losses in this case will be severe. We stop the homogenizer for reloading of the sample chamber when there is ~1 pulse of buffer/sample remaining in the sample chamber.
13. Dispense the Ni-IDA beads directly into Resuspension Buffer in a fume hood in order to minimize aerosolization of the beads.
14. It is advisable to take samples from every stage of the purification (cells before induction, induced cells, clarified lysate, *etc.*) to aid in troubleshooting should the need arise.
15. The most robust way to assess this is to do small-scale incubations with Ni-IDA beads to determine the fraction of protein that can be bound and sedimented in each case
16. We obtain ~1 ml of ~150 μM (~6.5 mg/ml) purified Vps1 GTPase-BSE per 2 l starting culture.

Acknowledgement

We thank Tyler M. Brady and Bryan A. Tornabene for assistance with protein purification. This work was supported by the National Institutes of Health grant GM120102 (M. G. J. Ford).

References

1. Ford MGJ, Chappie JS (2019) The structural biology of the dynamin-related proteins: new insights into a diverse, multitalented family. Traffic 20(10):717–740
2. Praefcke GJ, McMahon HT (2004) The dynamin superfamily: universal membrane tubulation and fission molecules? Nat Rev Mol Cell Biol 5(2):133–147
3. Hoppins S, Lackner L, Nunnari J (2007) The machines that divide and fuse mitochondria. Annu Rev Biochem 76:751–780
4. Antonny B, Burd C, De Camilli P et al (2016) Membrane fission by dynamin: what we know and what we need to know. EMBO J 35 (21):2270–2284
5. Faelber K, Gao S, Held M et al (2013) Oligomerization of dynamin superfamily proteins in health and disease. Prog Mol Biol Transl Sci 117:411–443
6. Varlakhanova NV, Alvarez FJD, Brady TM et al (2018) Structures of the fungal dynamin-related protein Vps1 reveal a unique, open helical architecture. J Cell Biol 217 (10):3608–3624
7. Arlt H, Reggiori F, Ungermann C (2015) Retromer and the dynamin Vps1 cooperate in the retrieval of transmembrane proteins from vacuoles. J Cell Sci 128(4):645–655
8. Chi RJ, Liu J, West M, Wang J, Odorizzi G, Burd CG (2014) Fission of SNX-BAR-coated endosomal retrograde transport carriers is promoted by the dynamin-related protein Vps1. J Cell Biol 204(5):793–806
9. Hayden J, Williams M, Granich A et al (2013) Vps1 in the late endosome-to-vacuole traffic. J Biosci 38(1):73–83
10. Kuravi K, Nagotu S, Krikken AM et al (2006) Dynamin-related proteins Vps1p and Dnm1p control peroxisome abundance in Saccharomyces cerevisiae. J Cell Sci 119 (Pt 19):3994–4001
11. Peters C, Baars TL, Buhler S, Mayer A (2004) Mutual control of membrane fission and fusion proteins. Cell 119(5):667–678
12. Smaczynska-de R II, Allwood EG, Aghamohammadzadeh S, Hettema EH, Goldberg MW, Ayscough KR (2010) A role for the dynamin-like protein Vps1 during endocytosis in yeast. J Cell Sci 123(Pt 20):3496–3506

Part II

Mammalian DSPs: Isolation and Biochemical Characterization

Chapter 3

A Single Common Protocol for the Expression and Purification of Soluble Mammalian DSPs from *Escherichia coli*

Natalia Stepanyants, Patrick J. Macdonald, Pooja Madan Mohan, and Rajesh Ramachandran

Abstract

Mammalian DSPs have been historically isolated either from native tissue sources or from transfected insect cell cultures via time-consuming and cumbersome protocols often yielding protein of variable quality and quantity. A facile and highly reproducible alternative methodology involving the heterologous expression and purification of soluble mammalian DSPs from *E. coli*, which yields highly active and functional protein of a uniform quality and quantity, free of spurious posttranslational modifications inherent to mammalian and insect cell expression systems, is described in this chapter.

Key words DSP Purification, *Escherichia coli*, Polyhistidine Tag, Dynamin, Drp1, OPA1

1 Introduction

The production of soluble, full-length mammalian DSPs, such as dynamin and dynamin-related protein 1 (Drp1), for biochemical, biophysical, and structural studies has customarily involved isolation from limiting native tissue sources (e.g., rat brain) [1, 2] or from cost-prohibitive baculovirus-infected or transfected insect cell culture expression systems [3–5]. These procedures generally entail long and unwieldy protocols that consume several days, if not weeks, for sufficient protein production. In the case of dynamin, purification has traditionally also entailed the secondary expression, purification, and bead adsorption of a protein partner (e.g., GST-tagged amphiphysin II SH3 domain) for affinity chromatography [1]. Moreover, mammalian proteins overexpressed in heterologous insect or yeast cell culture systems [6] are susceptible to

Natalia Stepanyants and Patrick J. Macdonald contributed equally to this work.

Rajesh Ramachandran (ed.), *Dynamin Superfamily GTPases: Methods and Protocols*, Methods in Molecular Biology, vol. 2159, https://doi.org/10.1007/978-1-0716-0676-6_3, © Springer Science+Business Media, LLC, part of Springer Nature 2020

spurious posttranslational modifications (e.g., phosphorylation artifacts) that likely cause protein heterogeneity and ultimately affect results. Phosphorylation is a known regulator of DSP interactions [7, 8]. Using human dynamin 1, Drp1, and short OPA1 (s-OPA1) as examples, we in this chapter report a relatively easy, quick, and cost-effective common protocol that combines inducible *E. coli* overexpression to FPLC-free, gravity flow-based tandem immobilized metal–ion affinity (IMAC) and ion exchange (IEX) chromatographies for both high-quality and high-quantity DSP production.

2 Materials

2.1 Plasmids for Protein Expression in E. coli

The gene (or construct) for the soluble DSP of interest must be subcloned in a plasmid vector (e.g., pRSET (Thermo Fisher) or pET (Novagen) vectors) that drives gene expression under the control of the strong bacteriophage T7 promoter. In addition, the DNA open reading frame (insert) of the gene of interest should be positioned in frame with a vector sequence that encodes either an N- or a C-terminal polyhistidine (6× His) fusion tag for IMAC purification. Further, genetically engineered BL21 (DE3) *E. coli* strains that encode for isopropyl-ß-D-thiogalactopyranoside-(IPTG)-inducible T7 RNA polymerase should be used for protein expression. We recommend the BL21 Star™ (DE3) *E. coli* strain (Thermo Fisher Scientific) that offers enhanced mRNA stability and relatively high-protein yield [9, 10]. Other BL21 (DE3) strains (e.g., pLysS) may also be used if the protein construct of interest appears to be toxic to bacterial growth. In this chapter, we use full-length versions of dynamin 1 (pET28a) and dynamin-related protein 1 (pRSET C) [10], as well as, s-OPA1 (ΔN195; pET28a) [11] to demonstrate the efficacy of our protocol.

2.2 Plasmid Transformation

1. One Shot® BL21 Star™ (DE3) Chemically Competent *E. coli*. (Thermo Fisher Scientific).
2. Incubator for growth of bacterial colonies on solid media set to 37 °C.
3. Water bath set to 42 °C.
4. Lysogeny broth (LB) agar bacterial culture plates composed of 1% (w/v) yeast extract, 0.5% (w/v) tryptone, 1% (w/v) NaCl, and 1.5% (w/v) agar with appropriate antibiotic (100 μg/mL ampicillin or 50 μg/mL kanamycin).

2.3 Expression

1. Sterile 250 mL culture flask.
2. Sterile 6 L culture flask.

3. Lysogeny broth (LB) culture medium: 1% (w/v) yeast extract, 0.5% (w/v) tryptone, and 1% (w/v) NaCl. Prepare 2 L and split into 1.9 L main (in 6 L flask) and 100 mL pilot (in 250 mL flask) fractions and autoclave.
4. 1000× Antibiotic stock solutions. Dissolve antibiotic in nuclease-free, sterile-filtered distilled water. Freeze in 1 mL aliquots at −20 °C until use. For examples described in this step, prepare 2 mL of each of the following:
   (a) 50 mg/mL kanamycin (e.g., pET28a).
   (b) 100 mg/mL ampicillin (e.g., pRSET C).
5. 500 mM IPTG. Prepare 2 mL in sterile-filtered water.
6. Incubator Shaker (with refrigeration).
7. Cell density meter.

2.4 Harvest and Lysis

1. Resuspension buffer: 20 mM HEPES set at pH 7.5, 500 mM KCl, 10 mM imidazole (from 1 M imidazole stock set at pH 8.0), and 1 mM 4-(2-aminoethyl) benzenesulfonyl fluoride hydrochloride (AEBSF). Add AEBSF to buffer just before use. Prepare 50 mL.
2. cOmplete™ EDTA-free Protease Inhibitor Cocktail Tablet (Roche).
3. Lysozyme powder.
4. Sonicator with a 0.5 in. tip diameter disruptor horn.
5. Ultracentrifuge with rotor and tubes.

2.5 Primary Purification by IMAC

1. 1 mL HisPur™ Ni-NTA Resin (Thermo Scientific).
2. Gravity-flow chromatography plastic column (10 mL capacity; we recommend the Bio-Rad Poly-Prep® Chromatography Column; *see* **Note 1**).
3. Wash buffer: 20 mM HEPES set at pH 7.5, 500 mM KCl, and 50 mM imidazole (from 1 M imidazole stock set at pH 8.0). Prepare 100 mL.
4. Elution buffer: 20 mM HEPES set at pH 7.5, 500 mM KCl, and 250 mM imidazole (from 1 M imidazole stock set at pH 8.0). Prepare 5 mL.
5. 1 M DTT. Prepare fresh and freeze in 100 μL aliquots at −20 °C until use.
6. SDS-PAGE apparatus.
7. Precast or hand cast 12% acrylamide gels; a total of two is required.
8. 5× SDS Sample Loading Buffer: 250 mM Tris HCl set at pH 6.8, 10 mM DTT, 10% (w/v) SDS, 0.05% (w/v) bromophenol

blue, and 30% (v/v) glycerol. Add 1 μL of β-mercaptoethanol per 5 μL of loading buffer when running.

9. SDS running buffer: 250 mM glycine, 25 mM Tris base, and 0.1% (w/v) SDS.
10. Coomassie Destain and Stain: 45% (v/v) methanol and 10% (v/v) acetic acid. For stain, add 0.25% (w/v) Coomassie Brilliant Blue G-250.
11. Slide-A-Lyzer™ Dialysis Cassettes, 10K MWCO, 0.5–3 mL capacity (Thermo Scientific) (*see* **Note 2**).
12. 18 Gauge hypodermic needle and 3 mL syringe.
13. Dialysis buffer 1: 20 mM HEPES at pH 7.5, 150 mM KCl, 1 mM DTT (add fresh), and 1 mM EDTA. Prepare 1 L.
14. Dialysis buffer 2: 20 mM HEPES at pH 7.5, 150 mM KCl, and 1 mM DTT (add fresh). Prepare 1 L (*see* **Note 3**).

2.6 Secondary Purification by IEX

1. 1.25 mL DEAE anion exchange resin (80% slurry, GE Healthcare).
2. Gravity-flow chromatography plastic column (10 mL capacity).
3. HEPES Buffered Saline (HBS): 20 mM HEPES at pH 7.5 and 150 mM KCl. Prepare 50 mL.
4. 2 M NaCl. Prepare 20 mL.
5. Glycerol, autoclaved.
6. 1 M DTT. Prepare fresh and freeze in 100 μL aliquots at −20 °C until use.
7. UV–Vis spectrophotometer.
8. Quartz cuvette.
9. Centrifugal concentrator (We recommend Pall Microsep™ Advance Centrifugal Devices with Omega Membrane—30K MWCO) (*see* **Note 4**).
10. SDS-PAGE apparatus.

3 Methods

3.1 Transformation of BL21 (DE3) E. coli

1. Thaw a 50 μL aliquot of One Shot® BL21 Star™ (DE3) Chemically Competent *E. coli* on ice for 10 min.
2. Add ~25–50 ng of plasmid DNA to the cells. Gently flick the tube 4–5 times to mix the cells and DNA. Do not vortex.
3. Place the mixture on ice for 30 min.
4. Heat shock at 42 °C in a water bath for 30 s.
5. Place cells on ice for 5 min.
6. Warm selection plates with appropriate antibiotic to 37 °C.

7. Spread the cells onto a selection plate using a sterilized spreading tool.
8. Invert the plates and incubate at 37 °C overnight (14–20 h).

3.2 Expression

1. Pick a single colony from the selection plate using a sterile micropipette tip and inoculate 100 mL of autoclaved LB pilot medium containing appropriate antibiotics (50 μg/mL kanamycin for pET28a or 100 μg/mL ampicillin for pRSET C).
2. Grow pilot at 37 °C overnight with shaking at 225 rpm until visibly turbid (<12 h).
3. Prepare bacterial glycerol stock for long-term storage (optional).
   (a) Add 150 μL of glycerol to a 2 mL cryo tube.
   (b) Transfer 850 μL of bacterial culture to the cryo tube, and mix well by gentle vortexing.
   (c) Snap–freeze bacterial stock (containing 15% (v/v) glycerol) by rapid immersion in liquid nitrogen and store at −80 °C.
4. Add pilot to the main flask with 1.9 L of LB and appropriate antibiotic (50 μg/mL kanamycin for pET28a or 100 μg/mL ampicillin for pRSET C). Grow at 37 °C with shaking for ~4–5 h until cells reach OD_{600} 0.8–1.0.
5. Immerse culture flask in an ice bath for 30 min, with intermittent shaking, until cool. Ensure the culture volume is surrounded by ice for uniform cooling.
6. Induce expression by adding IPTG to a final concentration of 0.5 mM (2 mL of 500 mM IPTG stock to 2 L culture).
7. Grow at 16 °C with shaking at 225 rpm overnight (~16–18 h).

3.3 Harvest and Lysis

These steps should be performed on ice, unless otherwise noted.

1. Split culture into two 1 L centrifuge bottles, and spin at 6900 × *g* for 10 min at 4 °C.
2. Resuspend pellets in 25 mL of resuspension buffer and transfer to 50 mL conical tube. Flash–freeze with liquid nitrogen. Cells can either be stored at −80 °C or proceed to step 3.
3. To thaw, make 10–15 mL of resuspension buffer with 1 regular protease inhibitor tablet and add to frozen cells. The final volume of buffer and cells should be 50 mL.
4. Rotate or rock tube at room temperature for 30–40 min until completely thawed.
5. Add lysozyme to a final concentration of 1 mg/mL to the cell suspension, and rock at 4 °C for 30 min.

6. Transfer cells to a 100 mL plastic beaker. Lyse cells by sonication. Set duty cycle to 30% and output to the microtip limit. Sonicate 5× 20 pulses (each pulse is 1 s on, 1 s off) with 1 min intervals in between each set of pulses. These settings may vary between sonicators (*see* **Note 5**).
7. Ultracentrifuge lysate at ~184 000 × *g* for 1 h at 4 °C. The supernatant contains soluble protein.

3.4 Primary Purification by IMAC

All buffers should be maintained cold, and column chromatography should be performed in the cold room (4 °C).

1. During step 7 of 3.3, equilibrate 1 mL final bed volume of HisPur™ Ni-NTA Resin in wash buffer. Add wash buffer in a final volume of 25 mL to 2 mL of a 50% (v/v) resin slurry in a 50 mL conical tube. Gently mix and centrifuge suspended resin at 500 × *g* for 10 min at 4 °C. Carefully pour off the supernatant without disturbing the settled resin. Rinse and repeat three times total.
2. Add supernatant from the ultracentrifugation step to the washed and settled Ni-NTA resin from the previous step. Rock gently on a nutating mixer for 1 h at 4 °C.
3. Centrifuge protein-bound resin at 500 × *g* for 10 min at 4 °C. Pour off supernatant without losing resin. Resuspend resin in 10 mL of wash buffer.
4. Rinse gravity-flow column with one column volume (10 mL) of wash buffer. Pour resuspended protein-bound resin into the column, and let it flow through. Rinse remnant beads in conical tube with 10 mL of wash buffer, add to the column, and let it flow through. At the last drop, add 1 mL of wash buffer without disturbing resin, and let it flow through. At last drop, add 9 mL of wash buffer, and let it flow through.
5. Position column over 1.5 mL microcentrifuge tubes placed on ice. Elute protein from resin by sequentially adding 0.5 mL, 1.0 mL, 0.5 mL, and 1.0 mL of elution buffer, and collect each fraction in a separate microcentrifuge tube.
6. To confirm the presence of protein, take 20 μL of sample from each fraction, and add 5 μL of 5× SDS-PAGE Loading Buffer containing fresh β-mercaptoethanol. Boil samples for 5 min. Perform SDS-PAGE and Coomassie staining according to standard protocols (Fig. 1a).
7. Pool protein-containing fractions, and add DTT fresh to a final concentration of 1 mM.
8. Pre-wet a Slide-A-Lyzer™ Dialysis Cassette (10K MWCO) in dialysis buffer 1 for about 1 min. Load cassette with pooled protein (2–3 mL total volume) using a 3 mL syringe fitted with an 18 gauge hypodermic needle according to manufacturer's protocols. Dialyze in buffer 1 overnight (~16–18 h) at 4 °C.

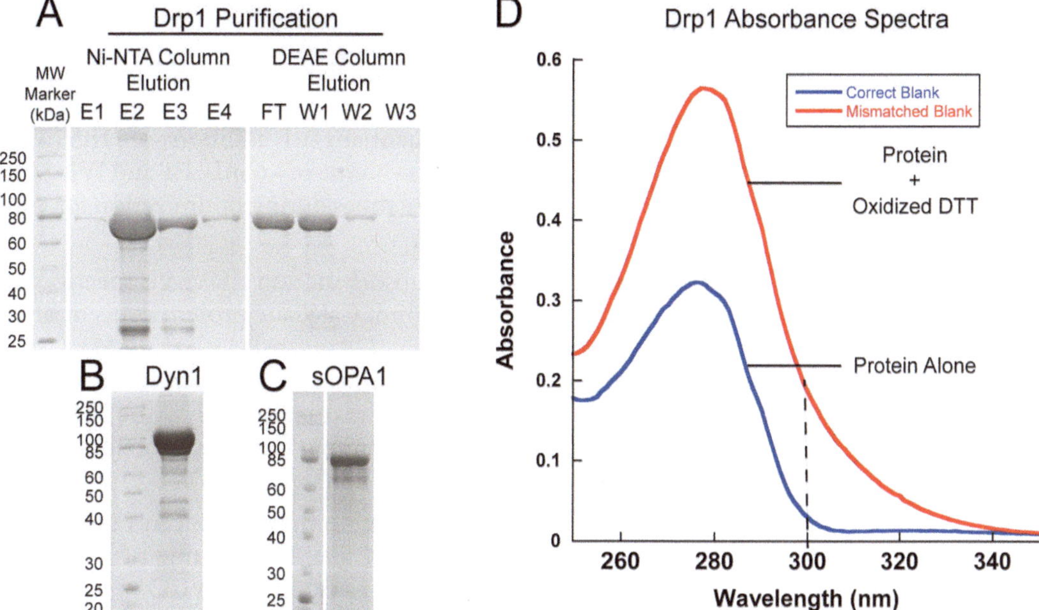

Fig. 1 (**a–c**) SDS-PAGE analyses of soluble DSP purification. (**a**) Drp1 elution fractions (E1-4) from primary purification by IMAC followed by secondary purification by IEX using a DEAE anion exchange resin (FT and W1-3). Drp1 is present in the FT and washes (W1-3), while DEAE retains most contaminants at the prescribed pH and salt concentration. (**b** and **c**) SDS-PAGE analyses of pooled and concentrated protein fractions of Dyn1 (B) and s-OPA1 (C) after DEAE IEX. (**d**) Absorbance spectra of purified Drp1 in buffer containing oxidized DTT recorded using a buffer-mismatched, DTT-free HBS blank (red trace) in comparison to the same recorded using a buffer-matched blank (blue trace) containing a 1:1 mixture of dialysis buffer 2 (containing the same amount of oxidized DTT) and HBS. Dashed vertical line shows a sloping edge beyond 300 nm for the blank-mismatched spectrum in comparison to a flat trace for the blank-matched spectrum. The blue trace also shows a small baseline offset at 340 nm and beyond, which may be subtracted from the protein absorbance value for baseline normalization

3.5 Secondary Purification by IEX

1. The next day, transfer the dialysis cassette from buffer 1 to buffer 2, and dialyze for an additional 2–4 h at 4 °C.
2. In the meantime, prepare a DEAE anion exchange column at room temperature by following these steps in sequence. Add 1.25 mL of an 80% (v/v) DEAE slurry to a 10 mL gravity-flow chromatography column. Rinse with 1 column volume (10 mL) of Milli-Q® Water. Charge resin with 1 column volume of 2 M NaCl. Rinse with 1 column volume of Milli-Q® Water. Equilibrate with 1 column volume of HBS. Do not let the resin dry out by retaining a fine layer of solution over the resin bed by capping the column exit in between steps (*see* **Note 1**).
3. Remove protein from dialysis cassette. Use same port as injection to avoid leakage of sample and the introduction of air bubbles.

4. Centrifuge protein sample at ~21,000 × *g* for 30 min at 4 °C to remove potential aggregates.
5. Add clarified protein to the equilibrated DEAE column, and collect flow through (FT) in a 15 mL conical tube immersed in ice, followed by three sequential 1 mL additions of HBS (wash (W) 1-3) for a total elution volume of ~6 mL. FT and W1-3 can also be collected in separate tubes during optimization for your DSP construct of interest.
6. Mix thoroughly and scan absorbance on a UV–Vis spectrophotometer from 200 to 800 nm. Calculate protein concentration using Beer–Lambert's law ($\{A_{280}/\varepsilon_{280}\}*10^6$ μM). Use a matching blank that corresponds to the buffer content of the protein solution. Use a 1:1 mixture of dialysis buffer 2 and HBS as blank if your pooled protein fractions containing 3 mL of flow-through + 3 mL of HBS (*see* **Note 6**).
7. Concentrate protein to a suitable storage concentration (generally between 10 and 20 μM final in a volume of ~1–2 mL) using a centrifugal device with a 30 kDa MWCO.
8. Determine final concentration by absorbance (*see* **Note 6**).
9. Add fresh DTT to the 1 mM final.
10. To assess purity, run sample on SDS-PAGE (Fig. 1b, c) (*see* **Notes 7** and **8**).
11. For storage, add glycerol to 10% (v/v) and mix gently but thoroughly by repeated pipetting. Adjust final concentration, aliquot protein samples to an appropriate volume (e.g., 20 aliquots of 100 μl each for 2 mL of concentrated protein at ~10 μM), flash-freeze in liquid nitrogen, and store at −80 °C.

4 Notes

1. Gravity-flow columns can be rinsed and reused, if promptly taken care of after use. Unattended columns generally suffer slower flow rates due to dried resin clogging the porous 30 μm polyethylene bed support that retains the resin in the column. Dispose of any such columns.
2. Cassettes are preferred over tubing for faster and more effective dialysis.
3. DTT should always be added fresh to buffer solutions.
4. We recommend use of the Pall Microsep™ Advance Centrifugal Devices for higher DSP recovery relative to other commercially available centrifugal devices.
5. Beaker containing resuspended bacterial cell lysate needs to be immersed in an ice bath during sonication to prevent heat

denaturation of proteins. The base of the ultrasonic tip should rest just above the bottom of the container to prevent frothing of the lysate and resultant protein denaturation.

6. An absorbance scan from 200 to 800 nm, instead of a point measurement at 280 nm, is highly recommended. Shape of the absorption spectrum provides information on potential blank-mismatch issues (Fig. 1d), as well as, on the presence of protein microaggregates that scatter light and artifactually increase protein absorbance values leading to erroneous concentration estimations. Aggregation-free protein samples (without a chromophore or a fluorophore moiety) generally do not absorb beyond 300 nm. Oxidized DTT (disulfide bond) absorbs UV light beyond 300 nm [12] and, at high enough concentrations in the protein sample, will add to protein absorbance, if not properly accounted for in the blank (which could be tricky) (Fig. 1d). Any residual, trailing (sloping) absorbance beyond 305 nm (as shown in Fig. 1d) is indicative of a potential blank buffer mismatch (mostly due to oxidized DTT present in the protein sample but not in the blank buffer), and if not, light scattering from protein microaggregates. Proteins that contain Trp and/or Tyr residues should have a well-defined absorbance peak positioned between 270 and 280 nm. Peak absorbance values should be between 0.1 and 1.0 (for a typical 1 cm pathlength cuvette) for accurate concentration measurements. Any trace, baseline offset (flat nonzero baseline 340 nm and above; Fig. 1d) may be subtracted from the protein A_{280} value for a more accurate protein concentration determination. We recommend rescanning samples at least three times to confirm an accurate read. Avoid introducing air bubbles into the cuvette, as this will cause an erroneous absorbance measurement.
7. An optional additional size-exclusion chromatography (SEC) step involving either a Superose 6 or Superdex 200 (GE Healthcare) column equilibrated with HBS may be incorporated at this stage for greater purification, especially for structural studies.
8. The polyhistidine tag may also be cleaved and removed using protease cut sites engineered between the fusion tag and protein of interest. Manufacturer's protocols compatible with DSP solubility are recommended for this step.

Acknowledgement

This work was supported by National Institutes of Health grant R01GM121583 awarded to R. R.

References

1. Stowell MH, Marks B, Wigge P, McMahon HT (1999) Nucleotide-dependent conformational changes in dynamin: evidence for a mechanochemical molecular spring. Nat Cell Biol 1:27–32
2. Wigge P, Kohler K, Vallis Y, Doyle CA, Owen D, Hunt SP, McMahon HT (1997) Amphiphysin heterodimers: potential role in clathrin-mediated endocytosis. Mol Biol Cell 8:2003–2015
3. Warnock DE, Hinshaw JE, Schmid SL (1996) Dynamin self-assembly stimulates its GTPase activity. J Biol Chem 271:22310–22314
4. Warnock DE, Terlecky LJ, Schmid SL (1995) Dynamin GTPase is stimulated by crosslinking through the C-terminal proline-rich domain. EMBO J 14:1322–1328
5. Damke H, Muhlberg AB, Sever S, Sholly S, Warnock DE, Schmid SL (2001) Expression, purification, and functional assays for self-association of dynamin-1. Methods Enzymol 329:447–457
6. Koirala S, Guo Q, Kalia R, Bui HT, Eckert DM, Frost A, Shaw JM (2013) Interchangeable adaptors regulate mitochondrial dynamin assembly for membrane scission. Proc Natl Acad Sci U S A 110:E1342–E1351
7. Anggono V, Robinson PJ (2007) Syndapin I and endophilin I bind overlapping proline-rich regions of dynamin I: role in synaptic vesicle endocytosis. J Neurochem 102:931–943
8. Cribbs JT, Strack S (2009) Functional characterization of phosphorylation sites in dynamin-related protein 1. Methods Enzymol 457:231–253
9. Mehrotra N, Nichols J, Ramachandran R (2014) Alternate pleckstrin homology domain orientations regulate dynamin-catalyzed membrane fission. Mol Biol Cell 25:879–890
10. Macdonald PJ, Stepanyants N, Mehrotra N, Mears JA, Qi X, Sesaki H, Ramachandran R (2014) A dimeric equilibrium intermediate nucleates Drp1 reassembly on mitochondrial membranes for fission. Mol Biol Cell 25:1905–1915
11. Ban T, Heymann JA, Song Z, Hinshaw JE, Chan DC (2010) OPA1 disease alleles causing dominant optic atrophy have defects in cardiolipin-stimulated GTP hydrolysis and membrane tubulation. Hum Mol Genet 19:2113–2122
12. Cleland WW (1964) Dithiothreitol, a New Protective Reagent for SH Groups. Biochemistry 3:480–482

Chapter 4

Affinity Purification and Functional Characterization of Dynamin-Related Protein 1

Ryan W. Clinton, Brianna L. Bauer, and Jason A. Mears

Abstract

Purification of dynamin-related proteins is complicated by their oligomeric tendencies. In this chapter, we describe an established purification regime to isolate the mitochondrial fission protein Drp1 using bacterial expression. Key attributes of dynamins include their ability to hydrolyze GTP and self-assemble into larger polymers under specific conditions. Therefore, the GTPase activity of Drp1 should be examined to confirm isolation of functional protein, and we describe a conventional colorimetric assay to assess enzyme activity. To determine the ability of Drp1 to self-assemble, we induce Drp1 polymerization through addition of a non-hydrolyzable GTP analogue. A sedimentation assay provides a quantitative measure of polymerization that complements a qualitative assessment through visualization of Drp1 oligomers using negative-stain electron microscopy (EM). Importantly, we highlight the caveats of affinity tags and the influence that these peptide sequences can have on Drp1 function given their proximity to functional domains.

Key words Dynamin-related protein, Mitochondrial fission, GTPase, Protein oligomerization, Electron microscopy

1 Introduction

Affinity purification is widely used to isolate recombinant proteins, and the breadth of tags reflects the variety of protein sequences with an attraction to a particular substrate. In classic studies, the inherent biochemical properties of dynamin were leveraged to isolate this protein from animal cells and tissues [1, 2]. More recently, several tags have been utilized to purify dynamin proteins with an emphasis on expressing proteins in bacterial culture to save on cost and to limit the time required for cell expansion before harvest [3–6]. In general, these efforts have been successful for both full-length and domain truncation constructs [7–10]. For full-length proteins, smaller affinity tags, including polyhistidine (6–10 amino acids) and calmodulin-binding (4 kDa) peptides (CBP), have been used to limit the amino acid length of these constructs, since bacteria struggle to express larger mammalian proteins. For smaller

Rajesh Ramachandran (ed.), *Dynamin Superfamily GTPases: Methods and Protocols*, Methods in Molecular Biology, vol. 2159, https://doi.org/10.1007/978-1-0716-0676-6_4,

domain constructs, larger tags, including maltose-binding peptides (MBPs, ~50 kDa), have been used to great effect. In this chapter, we will focus on the use of the calmodulin-binding peptide to isolate dynamin-related protein 1 (Drp1). This protocol yields pure protein through a one-step column purification. However, additional steps are required to remove the CBP tag and prevent any influence from this non-native peptide sequence on Drp1 function.

In fact, we have found that the CBP tag can alter protein behavior if not removed [11]. This is not surprising since both the N- and C-terminal ends of dynamins are adjacent to the GTPase and assembly domains in these proteins. Introducing an affinity peptide in this region has the potential to disrupt the assembly and activity of dynamins. In this chapter, we demonstrate examples where non-native tags impact both Drp1 self-assembly and activity. For this reason, we remove the CBP peptide using a protease cleavage site adjacent to the N-terminal methionine at the start of the Drp1 sequence. Subsequent purification using size-exclusion chromatography (SEC) is performed to isolate highly pure Drp1.

After purification, several experiments can be performed to assess the functionality of the protein. Fortunately, Drp1 is a GTPase, so this enzymatic activity can be measured [12]. Because Drp1 self-assembly leads to enhanced GTPase activity [13], this assay also provides an indication of polymer formation in the presence of specific substrates, including nucleotides and lipids. The extent to which the protein oligomerizes can be quantified using sedimentation methods, but negative-stain electron microscopy is essential for detecting the type of polymer being formed [14]. Dynamins have several interaction interfaces, which can lead to alternate modes of oligomerization, depending on the substrate used. Interestingly, affinity tags can alter the oligomeric tendencies of Drp1, and these effects are highlighted in the following sections:

2 Materials

Prepare all buffers (CalA, CalB, and SEC Buffer) fresh for same-day use and store at 4 °C. All solutions should be prepared using deionized water.

2.1 Purification of Drp1

1. CalA: 0.5 M L-arginine at pH 7.4, 0.3 M NaCl, 5 mM $MgCl_2$, 2 mM $CaCl_2$, 1 mM imidazole, and 10 mM β-Mercaptoethanol.
2. CalB: 0.5 M L-arginine at pH 7.4, 0.3 M NaCl, 2.5 mM EGTA, and 10 mM β-Mercaptoethanol.
3. SEC Buffer: 25 mM HEPES (KOH) at pH 7.5, 0.15 M KCl, 5 mM $MgCl_2$, and 10 mM β-Mercaptoethanol

* *Note: Filter the SEC Buffer before flowing it over the column in order to remove any undissolved particulates.*

4. Calmodulin Affinity Resin.
5. PreScission™ Protease.
6. Luria Broth (LB).
7. Amicon® Ultra-15 30 MWCO Centrifugal Filter.
8. Isopropyl-β-D-thiogalactopyranoside (IPTG).
9. Pefabloc® SC.
10. HiLoad 16/600 Superdex 200 Prep Grade.
11. BL21 Star™ (DE3) Chemically Competent *E. coli*.
12. pCal-n-Ek-Drp1 vector.
13. Ampicillin.
14. Coomassie.
15. Sonicator.
16. Ultracentrifuge.
17. Swinging Bucket Centrifuge.

2.2 Characterization of Drp1 GTPase Activity

1. Assembly Buffer: 25 mM HEPES (KOH) at pH 7.5, 150 mM KCl, and 10 mM β-mercaptoethanol.
2. Malachite Green Reagent: 1 mM Malachite Green Carbinol, 10 mM ammonium, and 1 N HCl (*see* **Note 1**).
3. Guanosine triphosphate (GTP).
4. $MgCl_2$.
5. Drp1 protein.
6. Plate reader.
7. MaxiSorp™ Nunc-Immuno™ Strips.
8. Heat block.
9. Potassium Phosphate Monobasic (KH_2PO_4).

2.3 Quantitative Assessment of Drp1 Oligomerization by Sedimentation Analysis

1. Purified protein in Assembly Buffer.
2. Methyleneguanosine 5′-triphosphate (GMPPCP).
3. Eppendorf 5424 Microcentrifuge.
4. Laemmli Buffer.
5. 4–20% SDS-PAGE Gel.
6. InstantBlue™ Coomassie Dye.
7. Gel imaging software (ImageJ or Gel Analyzer 2010, or alternative).

2.4 Qualitative assessment of Drp1 oligomerization by EM

1. Purified protein in Assembly Buffer.
2. Methyleneguanosine 5′-triphosphate (GMPPCP).
3. Carbon-coated TEM grids.
4. Parafilm.
5. Filter paper.
6. Uranyl acetate.
7. High-precision tweezers.
8. Electron Microscope—FEI Tecnai Spirit or F20.

3 Methods

3.1 Purification of Drp1

Both affinity and size-exclusion chromatography should be performed at 4 °C or in a cold room.

3.1.1 Preparation of Clarified Cell Lysate

1. Transform BL21 Star™ (DE3) Chemically Competent *E. coli* with the pCal- n-EK-Drp1 vector.
2. Inoculate 50 mL LB containing 100 g/mL ampicillin with a single colony of the BL21 Star™ (DE3) Cells containing the pCal- n-Ek-Drp1 plasmid. Grow this starter culture overnight at 37 °C with shaking at 200 rpm.
3. The following day, inoculate 2 L fresh LB containing 100 g/mL ampicillin with 40 mL of the starter culture (i.e., a 1:50 dilution with 20 mL into each 1 L). Incubate at 37 °C with shaking at 200 rpm until an OD_{600} of 0.6–0.8 is achieved (*see* **Note 2**).
4. Induce the cells with 1 mM IPTG, then shake at 200 rpm for 24 h at 18 °C.
5. Harvest the culture by centrifugation at 4300 × *g* for 20 min at 4 °C and discard the resulting supernatant (*see* **Note 3**).
6. Resuspend the resultant pellet 5 mL CalA per gram and then add 1 mM Pefabloc® SC.
7. Lyse the cells by sonication on ice. Sonicate at 95% amplitude for 10 s on and 5 s off for a total of 5 min of sonication. Repeat the sonication an additional four times, for a total of five sonications. Rest the cells on ice for a minimum of 2 min in between each sonication (*see* **Note 4**).
8. Pellet the cell debris by centrifugation at 150,000 × *g* for 1 h at 4 °C (*see* **Note 5**).
9. Collect the clarified cell lysate (CL), or supernatant, and store it on ice or at 4 °C.

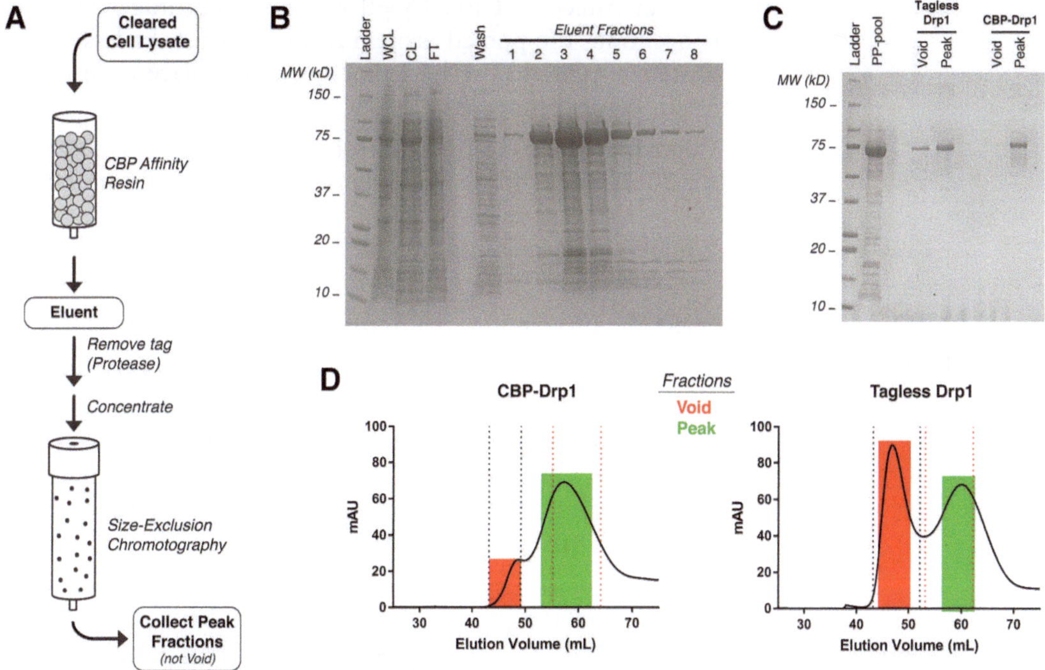

Fig. 1 Purification of Drp1. (**a**) Affinity purification followed by size-exclusion chromatography (SEC) is used to isolate purified protein. (**b**) A representative gel highlights the yield and purity of Drp1 following affinity purification. (**c** and **d**) Following SEC, void and peak fractions are observed, and most of the protein resides within the peak

3.1.2 Affinity Chromatography

1. Prepare the column with 4 mL Calmodulin Affinity Resin slurry (50% EtOH), which gives a column volume of 2 mL (*see* **Note 6**).
2. Wash the resin with at least 5 column volumes CalA (i.e., 10 mL for 2 mL resin).
3. Apply the CL to the column (Fig. 1a) and then allow it to flow through by gravity. Collect the flow-through (FT).
4. Wash the column with 50 mL CalA and discard the wash after it passes through the column.
5. Elute bound protein with CalB in eight 1 mL fractions (Fig. 1b).
6. Regenerate the column by first stripping it with 15 mL CalB and then re-equilibrating it with 15 mL CalA.
7. Repeat steps 12–15 by applying the FT to the column (*see* **Note 7**).
8. Pool all protein-containing fractions (*see* **Note 8**). To remove the affinity tag, add 2 L PreScission™ Protease per milliliter of pooled protein and then incubate overnight at 4°C (*see* **Note 9**).

9. Use an Amicon® Ultra-15 30 MWCO Centrifugal Filter to concentrate the pooled protein down to 5 mL or less. If the protein appears to precipitate or aggregate, place it on ice for 5 min before proceeding.
10. Centrifuge the concentrated protein for 10 min at 10,000 × *g* for 20 min at 4 °C. Separate the supernatant (*see* **Note 10**).

3.1.3 Size-Exclusion Chromatography

1. Equilibrate the column with one column volume of SEC Buffer (*see* **Note 11**).
2. Inject the protein into the 5 mL sample loop.
3. Separate the protein over the column at 0.5–1.0 mL/min (*see* **Note 12**). Do not exceed a pressure of 0.3 MPa.
4. After all the sample has gone through the column, two peaks should be observed (Fig. 1c, d). Collect the fractions corresponding to the second peak, as the first peak contains very little Drp1 (*see* **Note 13**).
5. Add glycerol as a cryoprotectant to a final concentration of 5%. Concentrate the protein down to 0.5–1.0 mL, once again, using a 30 MWCO concentrator.
6. Split the protein into 15–20 L aliquots. Flash-freeze these aliquots in liquid nitrogen and then store at −80 °C for later use (*see* **Note 14**).

3.2 Characterization of Drp1 GTPase Activity

1. Thaw all components, including GTP and Drp1, on ice.
2. Prepare the phosphate standard in Assembly Buffer using Potassium Phosphate Monobasic (KH_2PO_4) diluted to the following concentrations: 100, 80, 60, 40, 20, 10, 5, and 0 μm. Add 20 μL of the appropriate standard to the corresponding well.
3. Prepare a 3× GTP/$MgCl_2$ solution in Assembly Buffer, and store it on ice until use. The final concentrations of GTP/$MgCl_2$ should be 1 mM and 2 mM, respectively.
4. Prepare a 2.4× solution of Drp1 in Assembly Buffer, and store it on ice until use. The final concentration of Drp1 should be 500 nM.
5. Add 50 μL 2.4× Drp1 in triplicate to PCR tubes and then add 30 μL Assembly Buffer (*see* **Note 15**).
6. Move PCR tubes to a 37°C heat block and start the GTP hydrolysis reaction soon after by adding 40 μL 3× GTP/$MgCl_2$ with a multichannel pipette (*see* **Note 16**).
7. At desired time points, quickly remove 20 μL of the reaction mixture, and add it to a well containing 5 μL 0.5 M EDTA, which will quench the reaction (*see* **Note 17**).

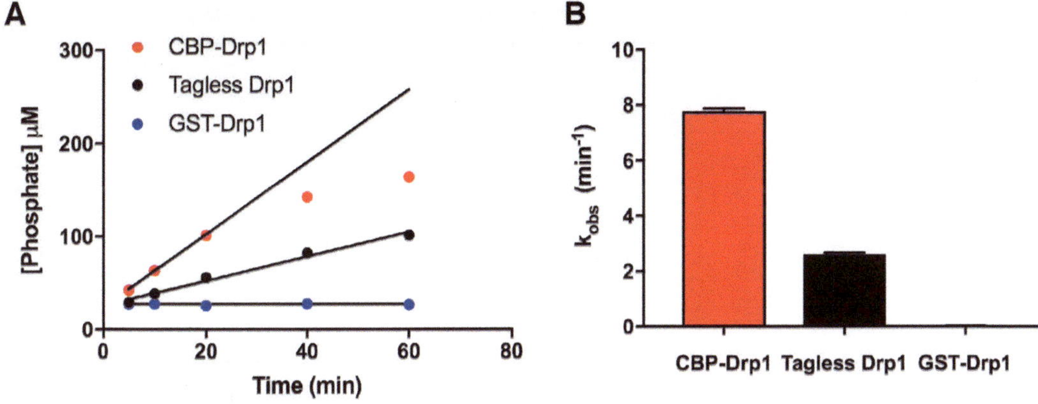

Fig. 2 Affinity tags impact Drp1 activity. (**a**) The GTPase activity of Drp1 with different affinity tags was measured using a colorometric assay that measures phosphate generation. (**b**) The calculated activity rates (k_{obs}) highlight the impact of distinct tags (CBP and GST) on Drp1 function

8. Add 150 μL Malachite Green Reagent to each well and then read the plate at A_{650} (*see* **Note 18**).
9. Plot the A_{650} values for the phosphate standards to create a standard curve.
10. Use this curve to calculate the amount of phosphate released at each time point (Fig. 2a). The linear portion of this graph can be used to calculate the amount of phosphate released (ΔY) over time (ΔX) at that Drp1 concentration (**Note 19**), which can be reported as the k_{obs} (Fig. 2b, *see* **Note 20**).

3.3 Quantitative Assessment of Drp1 Oligomerization by Sedimentation Analysis

1. Prepare 2× Drp1 in Assembly Buffer. The final concentration of Drp1 should be 2–5 M.
2. Prepare 2× GMPPCP/$MgCl_2$ in Assembly Buffer. The final concentrations of GMPPCP/$MgCl_2$ should be 1 mM and 2 mM, respectively (*see* **Note 21**).
3. Combine the 2× Drp1 and 2× GMPPCP/$MgCl_2$ in a 1:1 ratio (*see* **Note 22**).
4. Incubate at room temperature for 1 h (*see* **Note 23**).
5. Sediment via centrifugation at 16,100 × *g* for 30 min at 4 °C.
6. Collect the supernatant fractions without disturbing the pellets, and place in separate tube. Resuspend the pellet in ice-cold Assembly Buffer (*see* **Note 24**).
7. Add an appropriate amount of Laemmli Buffer to each sample before boiling for 10 min at 100 °C.
8. Run samples on a 4–20% SDS-PAGE Gel.
9. Stain the gel with InstantBlue™ Coomassie Dye (Fig. 3a).

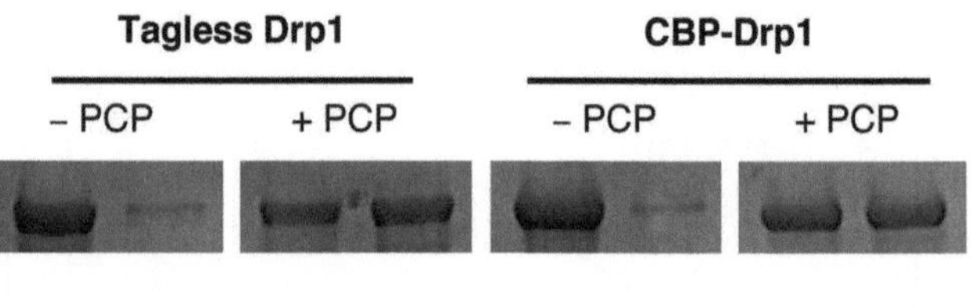

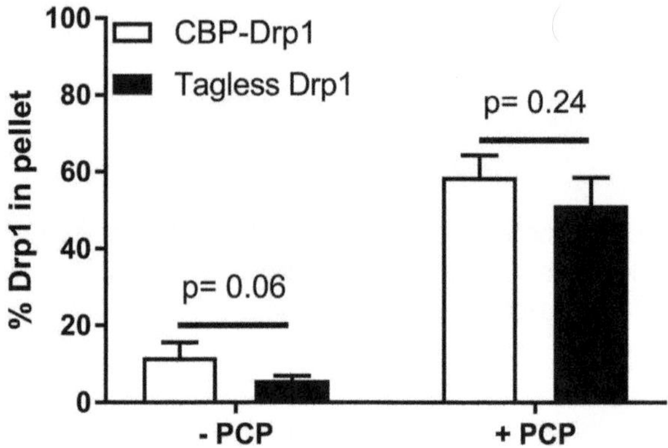

Fig. 3 Sedimentation assay can be used to quantify Drp1 oligomerization. (**a**) The sedimentation assay quantifies Drp1 self-assembly in the absence (−PCP) and presence (+PCP) of a non-hydrolyzable GTP analogue, GMPPCP. After centrifuging the sample, the supernatant and pellet fractions (left and right lanes, respectively) are run on a gel. (**b**) Imaging software was used to quantify the relative amount of Drp1 in the pellet

10. Quantify the relative amounts of Drp1 in each supernatant (S_{int}) and pellet (P_{int}) fraction using a gel imaging software, such as, ImageJ or GelAnalyzer2010 (*see* **Note 25**).
11. To calculate the percent of protein in the pellet (Fig. 3b), divide the measured pellet band intensity by the sum of the supernatant and pellet band intensities (i.e., $P_{int}/S_{int} + P_{int}$).

3.4 Qualitative Assessment of Drp1 Oligomerization by EM

1. Prepare samples containing 2 M Drp1 in Assembly Buffer (*see* **Note 26**).
2. Prepare carbon-coated mesh grids (*see* **Note 27**).
3. On a piece of parafilm, add 5–10 L of sample along with two drops of 2% uranyl acetate.
4. Place the grid, carbon-side down, onto the drop of protein sample. Allow the grid to remain on the drop for 30–60 s.
5. Blot the grid on a piece of filter paper to remove excess sample.
6. Wash the grid on the first drop of 2% uranyl acetate by touching the grid to the drop and then immediately blotting it on a piece of filter paper. Repeat wash.
7. Place the grid on the second 2% uranyl acetate for 30–60 s.

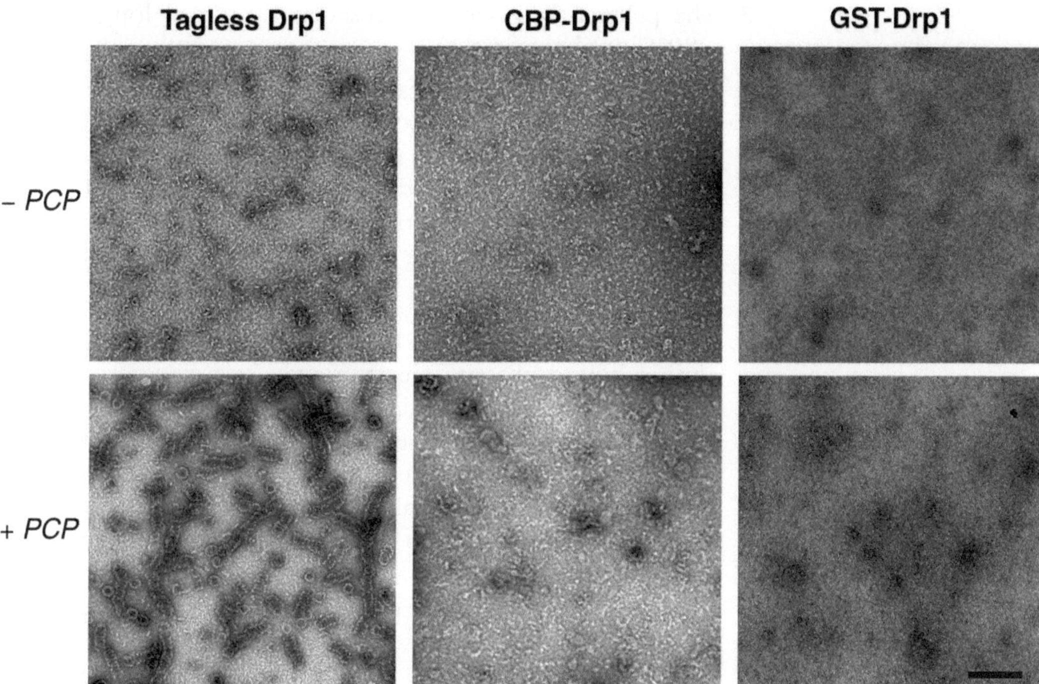

Fig. 4 Negative-stain election microscopy (EM) provides a qualitative assessment of Drp1 oligomerization. In the absence of nucleotide (top panels), Drp1 forms smaller multimers that appear as crescent-shaped particles. In the presence of GMPPCP (bottom panels), tagless Drp1 self-assembles into larger spiral structures. Both affinity tags (CBP and GST) limit the formation of functional assemblies. Scale bar, 100 nm

8. Remove excess stain by blotting the grid on a piece of filter paper.
9. An electron microscope is used to image the grid (*see* **Note 28**). Self-assembly can be assessed (Fig. 4), and morphological differences can be measured qualitatively and/or quantitatively (*see* **Note 29**). If the grid will not be visualized immediately, store it in a grid box under a vacuum until use.

4 Notes

1. To make the Malachite Green Reagent, begin by diluting the HCl to 1 N. Then, add the solid components and mix on a stir plate until fully incorporated into solution. Use a 0.45 mm filter to remove any remaining particles that did not dissolve the solution. Store the solution at room temperature in the dark.
2. We find that an appropriate OD is consistently obtained after 2.5 h of incubation.

3. The pelleted cells can be subsequently stored long-term at −80°C until purification.
4. Prior to sonication, the resuspended cells will appear cloudy, and after sonication, the solution should be noticeably more transparent. If the solution remains extremely cloudy, the sonication can be repeated an additional several times or 200 g/mL lysozyme can be added prior to sonicating.
5. Using a Beckman Type 45 Ti rotor, the centrifugation speed will be ~101,000 × *g*.
6. When adding resin to the column, washing the column, and eluting bound protein, pipette slowly in a circle around the periphery of the column in order to avoid channeling, and create and maintain an even layer of resin at the bottom of the column (alternatively, a frit can be used). Pipetting directly onto the resin can push it all to one side of the column, which could allow the protein to pass through the column without interacting with the resin.
7. We find that applying the FT to the column and repeating steps 12–15 twice works well.
8. The protein-containing fractions can be determined by SDS-PAGE or more quickly by taking 5 L from each fraction and adding 195 L Coomassie into corresponding wells.
9. If you do not remove the tag, the protein can behave differently. This is highlighted in later sections characterizing the activity and functional assembly of the Drp1 protein.
10. Traditionally, sedimentation assay has been performed using tabletop ultracentrifuges at higher speeds. We have found that a tabletop microcentrifuge is able to sediment Drp1 polymers to a similar extent.
11. We use a HiLoad 16/600 Superdex 200 Prep Grade column to purify Drp1 and equilibrate the column with one column volume, which is approximately 120 mL.
12. The fraction size we use is 1 mL.
13. The first peak may be intense because of higher scatter in the detector, but we observe little protein in these fractions (Fig. 1c, d). This partitioning of peaks is important to remove unwanted larger Drp1 aggregates.
14. If a specific stock concentration of protein is desired, an analytical assay to measure the concentration of protein can be performed prior to splitting it into aliquots. Then, the flow-through at the bottom of the concentrator from step 24 can be used to dilute the protein as needed.
15. Instead of adding 30 μL Assembly Buffer, 4× lipid or other cofactors in Assembly Buffer can be added to the Drp1 solution

and incubated for some period of time prior to the addition of GTP. Thus, the impact of lipids or partner proteins on the basal GTPase activity of Drp1 can be determined. When using a cofactor, we typically incubate it together with Drp1 for 15 min at room temperature before adding GTP.

16. We use a thermocycler set to hold its temperature at 37 °C indefinitely as a heat block.
17. For Drp1 without any lipids, partner proteins, or other cofactors added, the time points we normally use are 5, 10, 20, 40, and 60 min. In the presence of an agent that stimulates Drp1's GTPase activity, the time points should be scaled down appropriately. For example, when cardiolipin-containing liposomes are added, our time points are 2, 4, 6, 8, and 10 min.
18. The plate should be read within 10 min of adding the Malachite Green Reagent. This will help avoid excessive GTP hydrolysis due to the low pH of the Malachite Green Reagent.
19. The rates of GTP hydrolysis and K_{cat} (turnover number) of Drp1 are dependent on the concentration of protein used.
20. The presence of an affinity tag impacts the rate of hydrolysis by Drp1. In the case of the CBP tag, the rate is ~fourfold higher than the tagless protein (Fig. 2). Conversely, GST-tagged Drp1 has no apparent activity.
21. Also prepare 2× $MgCl_2$, without any GMPPCP, to serve as a control and measure sedimentation with the protein alone. As an alternative or in combination to GMPPCP, lipid nanotubes or liposomes can also be used to induce oligomerization and sedimentation.
22. It is recommended that the final volume is 20 L or greater. It is hard to siphon smaller volumes without agitating the pellet.
23. GMPPCP is a non-hydrolyzable GTP analogue. As an alternative, GTP and other analogues can be used. With GTP, the samples should be placed on ice after a defined length of incubation to stop the hydrolysis reaction. Centrifugation with this sample should be performed in a refrigerated microcentrifuge.
24. Be careful not to disturb the pellet while removing the supernatant as the pellet may be difficult to see or invisible. In a tabletop microcentrifuge, the pellet will typically form on one side of the microcentrifuge tube. We insert the tubes into the tabletop microcentrifuge in a uniform orientation and then pipette from the opposite side of the tube from the pellet, even though the pellet cannot be seen.
25. Oligomeric protein will be located in the pellet fraction. Note that the sedimentation values of CBP-tagged and tagless Drp1 are indistinguishable despite the difference in GTPase activities

(compare Figs. 2 and 3). This highlights the need for a multi-faceted characterization of purified protein.

26. To analyze the capability of the protein to form spirals, rather than its apo-state, 1 mM GMPPCP, a non-hydrolyzable GTP analogue, and 2 mM $MgCl_2$ can be included in the sample. Alternatively, nanotubes, liposomes, or partner proteins can also be incorporated to determine the ability of Drp1 to form polymers with these substrates. Incubate Drp1 together with these other components for 1–2 h at room temperature prior to imaging.
27. The grids may be glow discharged or plasma cleaned before the sample is applied in order to increase their hydrophilicity.
28. In the absence of Drp1 self-assembly, no discernible features are obvious. At higher magnification (~30,000–50,000×), a lawn of protein can be seen. Under conditions that promote Drp1 self-assembly, oligomers can be observed at lower magnification (~4000–6000×), and higher magnification imaging can identify general features of polymers (i.e., size, abundance, and helical patterns).
29. The differences between tagged (CBP and GST) and tagless protein are difficult to discern with the protein alone (*upper panels*, Fig. 4). However, addition of GMPPCP promotes protein oligomerization with nucleotide binding. With the tagless protein, spiral structures are readily observed and cover the EM grid (*lower left panel*). CBP-tagged Drp1 forms rings or arcs (*lower middle panel*), but fails to form extended polymers. GST-tagged protein is assembly incompetent as no polymers are observed (*lower right panel*). Therefore, the affinity tags clearly impact the assembly properties of Drp1.

Acknowledgement

The authors would like to acknowledge the Ramachandran lab for providing GST-tagged Drp1. Protein purification and characterization studies were performed by RWC and BLB. RWC was supported by the American Heart Association (16GRNT30950012). BLB is supported by the Molecular Therapeutics Training Grant (NIH, T32 GM008803-15). This work and JAM are supported by the NIH (R01 GM125844-01 and R01 CA208516-01A1).

References

1. Damke H, Muhlberg AB, Sever S et al (2001) Expression, purification, and functional assays for self-association of dynamin-1. Methods Enzymol 329:447–457
2. Quan A, Robinson PJ (2005) Rapid purification of native dynamin I and colorimetric GTPase assay. Methods Enzymol 404:556–569

3. Griparic L, van der Bliek AM (2005) Assay and properties of the mitochondrial dynamin related protein Opa1. Methods Enzymol 404:620–631
4. Ingerman E, Perkins EM, Marino M et al (2005) Dnm1 forms spirals that are structurally tailored to fit mitochondria. J Cell Biol 170:1021–1027
5. Wang L, Barylko B, Byers C et al (2010) Dynamin 2 mutants linked to centronuclear myopathies form abnormally stable polymers. J Biol Chem 285:22753–22757
6. Yoon Y, Pitts KR, McNiven MA (2001) Mammalian dynamin-like protein DLP1 tubulates membranes. Mol Biol Cell 12:2894–2905
7. Chappie JS, Acharya S, Liu YW et al (2009) An intramolecular signaling element that modulates dynamin function in vitro and in vivo. Mol Biol Cell 20:3561–3571
8. Chappie JS, Mears JA, Fang S et al (2011) A pseudoatomic model of the dynamin polymer identifies a hydrolysis-dependent powerstroke. Cell 147:209–222
9. Gao S, von der Malsburg A, Paeschke S et al (2010) Structural basis of oligomerization in the stalk region of dynamin-like MxA. Nature 465:502–506
10. Niemann HH, Knetsch ML, Scherer A et al (2001) Crystal structure of a dynamin GTPase domain in both nucleotide-free and GDP-bound forms. EMBO J 20:5813–5821
11. Francy CA, Alvarez FJ, Zhou L et al (2015) The mechanoenzymatic core of dynamin-related protein 1 comprises the minimal machinery required for membrane constriction. J Biol Chem 290:11692–11703
12. Leonard M, Song BD, Ramachandran R et al (2005) Robust colorimetric assays for dynamin's basal and stimulated GTPase activities. Methods Enzymol 404:490–503
13. Chang CR, Manlandro CM, Arnoult D et al (2010) A lethal de novo mutation in the middle domain of the dynamin-related GTPase Drp1 impairs higher order assembly and mitochondrial division. J Biol Chem 285:32494–32503
14. Mears JA, Hinshaw JE (2008) Visualization of dynamins. Methods Cell Biol 88:237–256

Chapter 5

Purification and Characterization of MxB

Frances Joan D. Alvarez and Peijun Zhang

Abstract

MxB/Mx2 is an interferon-induced dynamin-like GTPase, which restricts a number of life-threatening viruses. Because of its N-terminal region, predicted to be intrinsically disordered, and its propensity to self-oligomerize, purification of the full-length protein has not been successful in conventional *E. coli* expression systems. In this chapter, we describe an expression and purification procedure to obtain pure full-length wild-type MxB from suspension-adapted mammalian cells. We further describe how to characterize its GTPase activity and oligomerization function.

Key words Dynamin, Mx protein, GTPase, Mammalian expression, Purification, MBP fusion, Cryo-EM, Helical assembly, Negative-staining EM

1 Introduction

MxB/Mx2 is an interferon-stimulated Mx2 gene product, found in eukaryotes, that restricts viral pathogens, such as, human immunodeficiency virus (HIV) [1–3], herpesvirus [4], murine cytomegalovirus (MCMV) [5], equine infectious anemia virus (EIAV) [6], porcine reproductive and respiratory syndrome virus (PRRSV) [7], among others. Although the mechanistic details are still unclear, it is thought that MxB acts during early postentry stages of the infection by blocking viral genome replication.

MxB belongs to the dynamin family of large GTPases and, thus, shares the core architecture of its members [8–10]. The core is composed of a GTPase (G) domain, which catalyzes the hydrolysis of GTP; a stalk domain, which contains the major interfaces for homo-oligomerization; and a bundle-signaling element (BSE) domain, which connects the GTPase and stalk domains, thus transmitting conformational changes between the two domains. Unique to MxB is its N-terminal region (NTR), which contains a nuclear localization signal. For the anti-HIV activity of human MxB, its NTR and oligomerization properties are critical determinants, but the GTPase activity is dispensable. For equine MxB, a fragment of

Rajesh Ramachandran (ed.), *Dynamin Superfamily GTPases: Methods and Protocols*, Methods in Molecular Biology, vol. 2159, https://doi.org/10.1007/978-1-0716-0676-6_5, © Springer Science+Business Media, LLC, part of Springer Nature 2020

the NTR (1-25 aa) is not necessary, while the GTPase activity is important for restriction of EIAV [6]. In the case of herpesvirus restriction, both the NTR and the GTPase activity of MxB are important for its antiviral activity. These observations suggest a versatility of MxB in exerting different modes of inhibition for viral targets.

Structural analyses of human MxB have been hampered by the difficulty in purifying soluble full-length protein for in vitro studies [11]. The crystal structure of MxB was solved by removing its NTR and introducing stalk mutations, preventing higher-order oligomerization [9]. Clearly, study of the full-length protein with intact GTPase function and self-assembly properties is important to understanding its mode of action. Furthermore, modifications introduced into the crystallization constructs of MxB alters the MxB structure, particularly in regions involved in the oligomerization interfaces, as compared to the cryo-EM structure of full-length wild-type MxB [8]. In our laboratory, we exhausted various strategies of expressing and purifying the full-length wild-type protein in *E. coli.* We used a comprehensive list of expression constructs containing fusions and affinity tags and expressed them in various strains at different expression conditions, but none resulted in soluble protein, in amounts necessary for biochemical and biophysical studies. We then switched to baculovirus expression in insect cells and observed very good expression of MxB during our preliminary experiments. However, throughput is slow when testing different expression constructs and fusion tags for optimal protein production because it usually takes a few weeks for the generation and characterization of the viruses.

In this chapter, we present expression, purification, and characterization of recombinant full-length wild-type MxB in suspension-adapted mammalian cells.

2 Materials

Prepare all solutions using deionized water and analytical grade reagents. Prepare and store all purification reagents at 4 °C temperature (unless indicated otherwise). No sodium azide was added to the reagents.

2.1 Transfection

1. Plasmid DNA.
2. ExpiFectamine™ 293 Expression Kit (Thermo Fisher Scientific).
3. Expi293F™ cells (A14527, Thermo Fisher Scientific).
4. ExpiFectamine™ 293 Transfection Kit contains the ExpiFectamine™ 293 Reagent, Transfection Enhancer 1 and Transfection Enhancer 2 (A14524 Thermo Fisher Scientific).

5. Expi293™ Expression Medium (A14351, Thermo Fisher Scientific).
6. Opti-MEM™ I Reduced Serum Medium (31985-062, Thermo Fisher Scientific).
7. Sterile vent-cap Erlenmeyer flasks with baffled bottom.
8. Required infrastructure: basic cell culture room with Class II laminar flow hood and incubator with temperature and carbon dioxide control.
9. Equipment and reagents for cell counting and viability (light microscope, hemocytometer with trypan blue).

2.2 Purification

1. Phosphate-buffered saline (PBS).
2. HNG buffer (50 mM HEPES—KOH (pH 8), 250 mM NaCl, and 5% glycerol).
3. Amylose resin (New England Biolabs).
4. Type B pestle Dounce homogenizer (Kimble Chase).
5. Empty Chromatography Column (Bio-Rad).
6. Sephacryl® S500-HR (Sigma-Aldrich).
7. Concentrator with molecular weight cutoff (MWCO) of 50 kDa (Millipore).
8. Refrigerated table-top centrifuge (Eppendorf).

2.3 GTPase Assay

1. Spectrophotometer.
2. Quartz microcuvette.
3. GTPase reaction buffer (50 mM Hepes—KOH (pH 7), 150 mM NaCl, 10 mM $MgCl_2$, 2 mM dithiothreitol (DTT), 4 mM phosphoenolpyruvate (PEP), 0.35 mM NADH, 25 units of pyruvate kinase/lactate dehydrogenase, and 1 mM GTP).

2.4 Tube Assembly and ImmunoGold Labeling

1. 400 mesh carbon-coated electron microscopy grid (Ted Pella).
2. Anti-maltose-binding protein (MBP) (Abcam).
3. 5 nm gold-conjugated secondary antibody (Ted Pella).
4. Bovine serum albumin (BSA).
5. 2% uranyl formate.
6. Purified rhinovirus (HRV) 3C protease or PreScission protease.

3 Methods

3.1 Construction and Purification of Plasmid DNA

We used pcDNA3.1(+) (Invitrogen), which is a mammalian expression vector under the control of the cytomegalovirus (CMV) promoter and ampicillin for antibiotic selection in *E. coli*. We made the

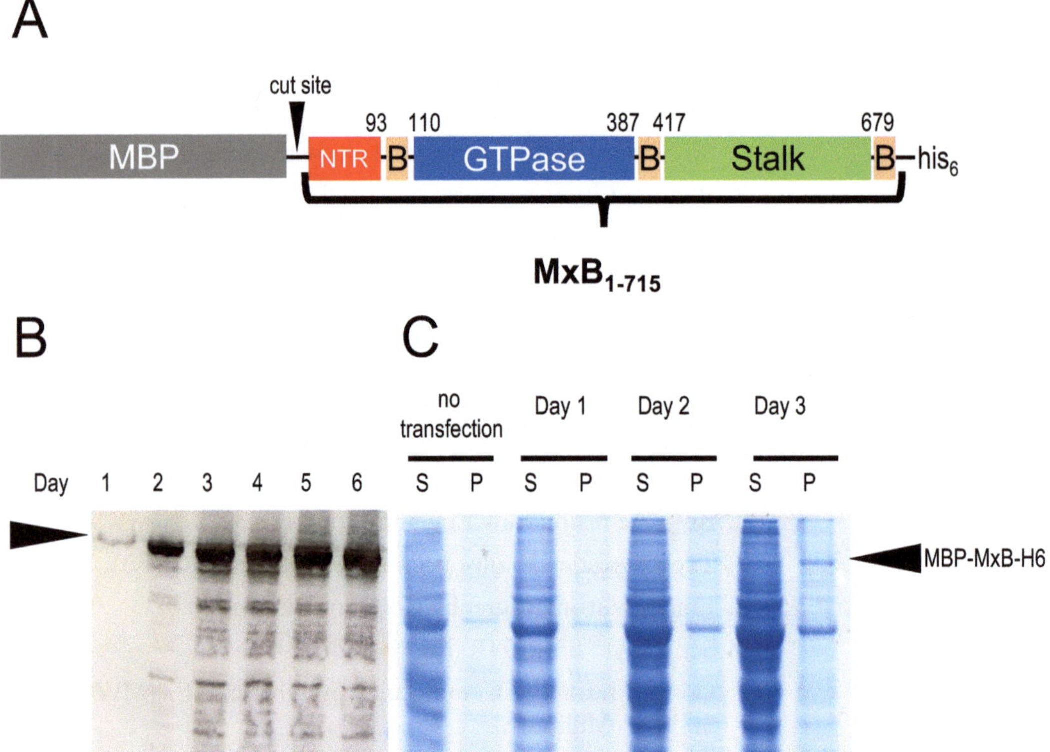

Fig. 1 (**a**) Expression construct containing the gene for full-length (1-715) wild-type human MxB, N-terminal maltose binding protein (MBP) fusion tag, and C-terminal hexahistidine tag. The schematic of the domain structure of MxB is shown: N-terminal region (NTR), BSE domain (B), GTPase domain, and stalk domain. (**b**) Time-course small-scale expression of MBP–MxB–H6 (black arrow) starting from day of transfection (day 1). A western blot using antibody against MBP is used to detect total expression of MBP–MxB–H6, which is increasing in amount during the course of expression. (**c**) Solubility of expressed protein is assayed by lysis of a small aliquot of cells followed by centrifugation to separate supernatant (S) and pellet (P), representing the soluble and insoluble fractions, respectively. The amount of MBP–MxB–H6 in the pellet increases during the course of expression

construct by Gibson Assembly to insert the expression construct (Fig. 1a) within the multiple cloning site (MCS) of the vector. The construct begins with a Kozak consensus sequence to initiate translation in eukaryotic cells, followed by the maltose-binding protein (MBP) tag, a linker containing the human rhinovirus (HRV) 3C protease (e.g., PreScission protease) cleavage site, the MxB gene, a hexahistidine tag, and stop codon. We designed tags on both ends to allow for multistep affinity purification, if necessary, and for monitoring the expression of a full, intact protein using antibodies against both tags (*see* **Note 1**).

The quality of plasmid DNA is critical to the success of transfection. We maintain the plasmid in *E. coli* DH5α cells and grow enough culture for a maxiprep purification (the volume will depend on the kit or procedure used). We tested different DNA purification kits and various protocols based on cesium chloride ($CsCl_2$) density-gradient purification. We found that using the NucleoBond Maxiprep kit (Takara Bio) was the most efficient method that yielded high-quality plasmid DNA, wherein purification is finished in a few hours and transfection can be performed on the same day (*see* **Note 2**).

3.2 Maintenance of Cells

Frozen Expi293F™ cells are thawed at 37 °C in a water bath and added to prewarmed expression media in a disposable sterile vent-cap Erlenmeyer flask with baffled bottom to achieve a concentration of 3–5 × 10^6 cells/ml (see more details from manufacturer's manual). The minimum volume should be 20% of the volume of the baffled flask to minimize foaming and maximum of 40% of the flask volume to allow for proper aeration of the cell suspension. The cells are monitored for growth and viability, ensuring that cells are not aggregated by visualization under the light microscope and at least 90–95% of cells are alive using trypan blue as indicator of dead cells. After a few passages, when cells are consistently at >95% viability, the cells can be used for transfection.

Because the cells are not maintained in the presence of antibiotics or antifungal reagents, aseptic technique must be practiced at every step to prevent contamination of the cells. Sterile single-use polycarbonate flasks also help in preventing contamination.

If an incubator with shaker, carbon dioxide, and humidity controls is not available, one can fit an orbital shaker, preferably with digital control for consistent agitation, in a standard cell culture incubator. Place a tray of distilled water at the bottom level to keep the incubator humidified (*see* **Note 3**).

3.3 Transfection

We essentially followed the manufacturer's instructions for using the transfection kit. Ensure that the cells will be at ~5 × 10^6 cells/ml on the day of transfection by splitting the cells in new media the day before (Note: doubling time is 24 h). On the day of transfection, prewarm all the media at 37 °C. Decontaminate all containers/equipment and perform all the steps inside a Class II laminar flow hood. Observe aseptic technique. For a small-scale culture:

1. Add 30 μg of plasmid DNA to 1.5 ml in Opti-MEM™ I Reduced Serum. Incubate for 5 min.
2. Add 80 μl of ExpiFectamine™ 293 Reagent to 1.5 ml in Opti-MEM™ I Reduced Serum. Incubate for 5 min.
3. Mix tubes containing DNA and ExpiFectamine™ 293 Reagent and incubate for 25 min.
4. Add the DNA/ExpiFectamine mixture to 25.5 ml of 2.9 × 10^6 cells/ml of Expi293F™ cells in a 125 ml Erlenmeyer flask. Final volume will be ~28.5 ml of ~2.5 × 10^6 cells/ml (25.5 ml cells + 3 ml of DNA/ExpiFectamine mixture).

5. Incubate at 37 °C and 125 rpm agitation with 8% CO_2 in air and in humidified environment.
6. After 18 h of incubation, add 150 μl Transfection Enhancer 1 and 1.5 ml of Transfection Enhancer 2 into the suspension.

Scale up the culture size following the DNA/transfection reagent/culture size above. Perform an expression and solubility test for new constructs to determine the optimum duration of transient expression. For MxB, we found that most of the expressed protein is soluble 24 h after addition of transfection enhancers (Fig. 1b, c).

3.4 Purification

Twenty-four hours after the addition of the enhancers, harvest cells by centrifugation at low speed (100 × *g*). Resuspend the cell pellet gently with cold phosphate-buffered saline to wash the cells. Spin the cells again at low speed, flash–freeze the cell pellet and store the pellet at −80 °C for later use.

3.4.1 Lysis and Collection of Soluble Fraction

1. Prepare the purification buffer (HNG): 50 mM Hepes—KOH (pH 8), 250 mM NaCl, and 5% glycerol (*see* **Note 4**).
2. Thaw and resuspend the frozen cell pellet in the HNG buffer supplemented with the following:

   1% Tween, 0.3% NP-40, 5 mM $MgCl_2$, protease inhibitors (1 tablet Roche per 50 ml), 50 μg/ml DNAse, and 2 mM DTT (*see* **Note 5**).
3. Incubate the cells in lysis buffer with rotation at 4 °C for 1 h. Then homogenize the lysate by 15 strokes in an ice-cold, tight-fitting Dounce homogenizer (type B pestle).
4. Centrifuge the homogenate at 21,000 × *g* at 4 °C for 30 min using a refrigerated table-top centrifuge.

3.4.2 Affinity Chromatography

1. After centrifugation, collect the supernatant and mix with amylose agarose resin (1 ml per 50 ml of cell suspension) pre-equilibrated with buffer HNG + 2 mM DTT.
2. Transfer the supernatant/resin mixture into a disposable column. Let flow through.
3. Wash unbound protein from the resin with at least 50× the volume of the resin. For some MxB mutants, a high salt (750 mM NaCl) wash is added as a step before completing the wash in 250 mM NaCl.
4. To elute the protein, plug the end of the column to stop the flow and add 3× resin volume buffer A + 100 mM maltose. Gently mix the resin + elution buffer using a clean plastic inoculation loop. Keep the column at 4 °C for 15 min before collecting the elution. At this point, ~90% of the protein is already eluted off the resin, but the elution step may be repeated to possibly elute more protein.

5. Combine and concentrate the elution volumes using concentrator with MWCO of 50 kDa up to 5 mg/ml, if necessary. Protein concentration is determined by measuring absorbance at 280 nm and using the calculated molecular weight and extinction coefficients 129,421.8 g/mol and 116,660 $M^{-1}cm^{-1}$, respectively.

3.4.3 Size-Exclusion Chromatography

Single-step affinity purification yields ~1 mg total protein per liter of culture media of fairly pure MxB. To further polish the purity, mainly to remove the excess MBP tag, we injected the protein into a Sephacryl® S500-HR (Sigma-Aldrich), which allows separation of a wide range of molecular weight of proteins (40–20,000 kDa) (*see* **Note 6**). MxB eluted into a single peak, but examination of the fractions under the peak showed that MxB exists in various oligomeric conformations (Fig. 2), which indicated its propensity to form helical tubes akin to the behavior of members within the dynamin superfamily [12].

3.5 Characterization

3.5.1 GTPase Activity

The ability of MxB to hydrolyze GTP was assessed using a continuous NADH- (reduced form of nicotinamide adenine dinucleotide) coupled assay (*see* **Note 7**) in a microcuvette format.

1. Set the spectrophotometer to read continuously for 30 min at 37 °C at 340 nm. Blank the machine with a quartz microcuvette containing deionized distilled water.
2. Prepare a 100 μl reaction mixture by mixing the reagents in a microcentrifuge tube to achieve the following final concentrations in the assay solution: 50 mM Hepes—KOH (pH 7), 150 mM NaCl, 10 mM $MgCl_2$, 2 mM dithiothreitol (DTT), 4 mM phosphoenolpyruvate (PEP), 0.35 mM NADH, 25 units of pyruvate kinase/lactate dehydrogenase, and 1 mM GTP. Warm the mixture at 37 °C in a heat block for a few minutes (*see* **Note 8**).
3. Add 1.5 μM MxB into the reaction mixture and quickly place 90 μl of the mixture into the microcuvette for measurement. For negative control, record measurements using the purification buffer only, in place of the protein (*see* **Note 9**).
4. Record the decrease in the NADH absorbance at 340 nm over time. Subtract the reading from the negative control to correct for background. The rate of NADH oxidation is taken as the slope of the linear portion of the resulting plot and is directly proportional to the amount of GTP hydrolyzed (Fig. 3a).

3.5.2 Tubular Assembly

The assembly of MxB into tubes is initiated by diluting the salt in the sample buffer to 150 mM NaCl or less, for at least an hour at room temperature. MxB tubes grow longer with extended incubation (overnight to a few days) (Fig. 3b).

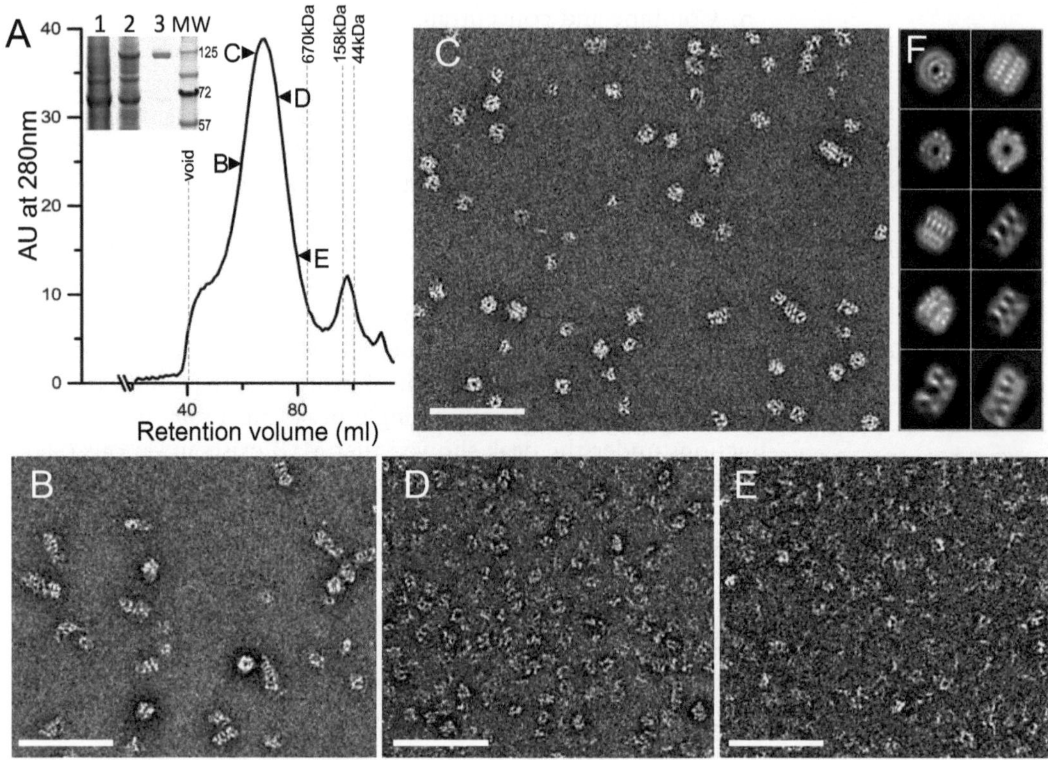

Fig. 2 Full-length, wild-type MxB purified as oligomers. (**a**) Purification of MBP–MxB–H6 from Expi293™ cells by amylose affinity chromatography, followed by gel filtration through Sephacryl® S500-HR column. The fractions indicated by arrowheads were visualized by negative-stain EM. Inset, Coomassie-stained SDS-PAGE gel of untransfected cells (1), transfected cells (2), and elution from amylose resin (3). Molecular weight (MW) markers are shown in kDa. (**b–e**) Representative micrographs by negative-stain EM of the indicated fractions in (**a**). (**f**) 2-D class averages of the negative-stained MxB sample from the fraction "C". Scale bars, 200 nm. Reproduced from [8], with permission from Sci Adv [8]

To determine the position of the NTR of MxB in its helical form, we performed on-grid immunogold labeling of MBP visualized by negative-stain electron microscopy (EM) as indirect localization of the NTR. We observed decoration of gold particles on the surface of the tubes, indicating that the NTR of MxB is oriented on the surface of the tubes (Fig. 3c).

1. Prepare MxB tubes as described above. Apply 3.5 μl of the sample on a 400 mesh carbon-coated EM grid. Incubate for 1 min and blot the rest of the liquid using a Whatman® #1 filter paper.
2. Float the grid on the following solutions at 4 °C (*see* **Note 10**):
   (a) Blocking buffer [bovine serum albumin (BSA; 10 mg/ml) in oligomerization buffer] for 5 min, twice.
   (b) Blocking buffer containing primary antibody against MBP tag (Abcam) (1:250 dilution) for 1 h.

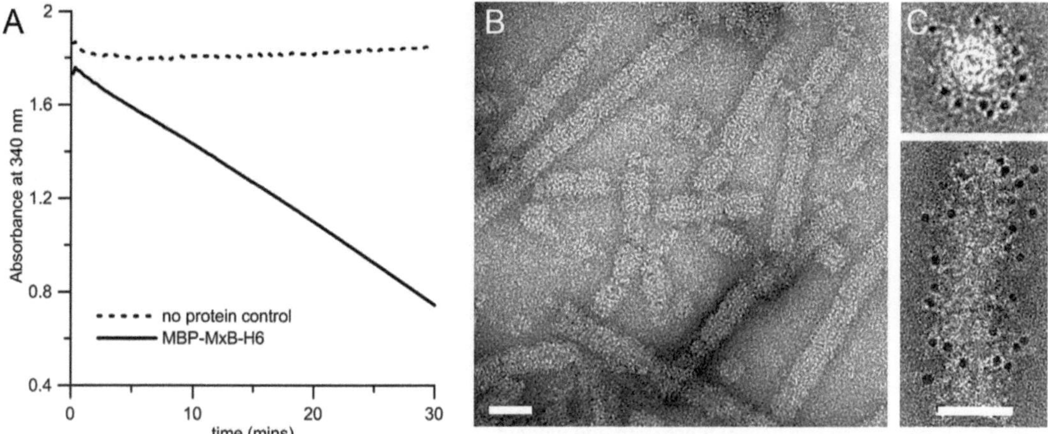

Fig. 3 MxB GTPase activity and tubular assembly. (**a**) The decrease in the NADH absorbance at 340 nm over time. (**b**) MBP–MxB self-assembled into tubular structures at 150 mM NaCl. (**c**) Gold-labeling of MBP shows that MBP is located on the surface of assembled MBP–MxB tubes. End on view (top) and side view (bottom) are shown. Scale bars, 50 nm in **b** and **c**. Reproduced from [8], with permission from Sci Adv [8]

(c) Blocking buffer for 5 min, twice.

(d) Blocking buffer containing a 5 mm gold-labeled secondary antibody (Ted Pella) (1:250 dilution) for 1 h.

(e) Blocking buffer for 5 min.

(f) Oligomerization buffer for 1 min, twice.

(g) 2% Uranyl formate for 1 min (*see* **Note 11**).

3. Blot the grid with filter paper and let the grid completely dry before imaging on the electron microscope.

3.5.3 Cleavage of MBP Tag

The MBP tag is cleaved from MxB by adding purified HRV3C (also known as PreScission protease) in 1:10 (protease to MxB molar ratio) to MxB in the purification buffer at 4 °C for 1 h. Monitor the completeness of the reaction by running an SDS-PAGE, followed by western blots using antibodies against MxB or MBP (Fig. 4a). We found that cleavage of MBP from MxB induced tube formation of MxB even at 250 mM NaCl, but those tubes are bundled together (Fig. 4b, c).

4 Notes

1. We made a similar construct using glutathione S-transferase (GST) tag in place of MBP, but the protein did not express well, which highlights the importance of testing different tags and constructs for optimum expression.
2. Ensure that plasmid DNA is fully dried and free of ethanol prior to resuspension in sterile double distilled water.

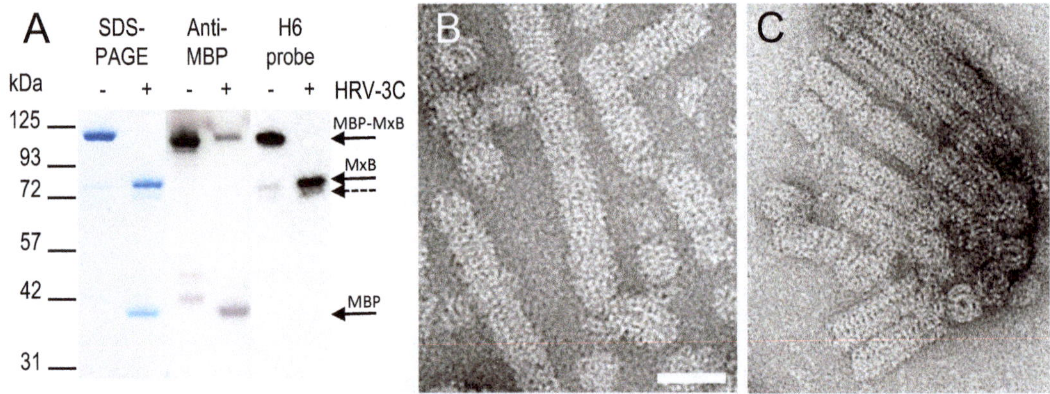

Fig. 4 MxB assembles into helical tubes with and without MBP tag. (**a**) Coomassie-stained SDS-PAGE gel and the corresponding western blots, with the indicated primary antibody or probe, of MBP–MxB–H6 without or with HRV 3C at 150 mM NaCl. Protein bands are indicated by arrows on the right. Dashed arrow points to a minor species of MxB with both MBP and the linker between MBP and MxB removed. (**b** and **c**) Negative-stain EM images of samples in (**a**) without the protease (**b**) and with the protease (**c**). Removal of MBP tag resulted in ordered MxB tubes but largely aggregated. Scale bar, 50 nm. Reproduced from [8], with permission from Sci Adv [8]

3. We attempted using a glass spinner culture bottle with a magnetic stir plate to maintain the cells, but, in our hands, the cell viability was consistently low. Reagents used for cleaning the glass or the heat generated by the spinning of the magnetic stirrer may account for low cell viability.
4. We performed small-scale purification using higher concentrations of NaCl in the purification buffer, but the amount of protein bound to resin also decreased.
5. Avoid concentrating the cells more than five times (i.e., for 30 ml cells worth of cell pellet, resuspend with no less than 6 ml of lysis buffer) to prevent aggregation or higher-order oligomerization.
6. Affinity-purified MxB elutes in the void volume of a Superdex 200 (GE Healthcare) column.
7. The recipe used in this study is modified for cuvette measurements based on a plate assay by Ingerman and Nunnari [13].
8. A 5× concentrated stock of the reaction mixture can be made ahead of time, frozen in −20 °C and kept away from light (NADH is light sensitive). The source of potassium needed in the assay is in the form of KOH used to adjust the pH of HEPES. If HEPES solution is prepared another way, supplement the reaction mixture with KCl to a final concentration of 7.5 mM.

9. The volume of the MxB added into the reaction mixture should be kept small enough to minimize the effect of NaCl carried from the purification buffer to assay conditions.
10. For negative control, prepare another grid without applying MxB tubes.
11. Uranyl formate is classified as a radioactive material and appropriate handling must be observed.

References

1. Goujon C, Moncorge O, Bauby H, Doyle T, Ward CC, Schaller T, Hue S, Barclay WS, Schulz R, Malim MH (2013) Human MX2 is an interferon-induced post-entry inhibitor of HIV-1 infection. Nature 502(7472):559–562
2. Kane M, Yadav SS, Bitzegeio J, Kutluay SB, Zang T, Wilson SJ, Schoggins JW, Rice CM, Yamashita M, Hatziioannou T, Bieniasz PD (2013) MX2 is an interferon-induced inhibitor of HIV-1 infection. Nature 502 (7472):563–566
3. Liu Z, Pan Q, Ding S, Qian J, Xu F, Zhou J, Cen S, Guo F, Liang C (2013) The interferon-inducible MxB protein inhibits HIV-1 infection. Cell Host Microbe 14(4):398–410
4. Schilling M, Bulli L, Weigang S, Graf L, Naumann S, Patzina C, Wagner V, Bauersfeld L, Goujon C, Hengel H, Halenius A, Ruzsics Z, Schaller T, Kochs G (2018) Human MxB protein is a Pan-herpesvirus restriction factor. J Virol 92 (17):e01056–e01018
5. Jaguva Vasudevan AA, Bahr A, Grothmann R, Singer A, Haussinger D, Zimmermann A, Munk C (2018) MXB inhibits murine cytomegalovirus. Virology 522:158–167
6. Meier K, Jaguva Vasudevan AA, Zhang Z, Bahr A, Kochs G, Haussinger D, Munk C (2018) Equine MX2 is a restriction factor of equine infectious anemia virus (EIAV). Virology 523:52–63
7. Wang H, Bai J, Fan B, Li Y, Zhang Q, Jiang P (2016) The interferon-induced Mx2 inhibits porcine reproductive and respiratory syndrome virus replication. J Interferon Cytokine Res 36 (2):129–139
8. Alvarez FJD, He S, Perilla JR, Jang S, Schulten K, Engelman AN, Scheres SHW, Zhang P (2017) CryoEM structure of MxB reveals a novel oligomerization interface critical for HIV restriction. Sci Adv 3(9):e1701264
9. Fribourgh JL, Nguyen HC, Matreyek KA, Alvarez FJD, Summers BJ, Dewdney TG, Aiken C, Zhang P, Engelman A, Xiong Y (2014) Structural insight into HIV-1 restriction by MxB. Cell Host Microbe 16 (5):627–638
10. Haller O, Staeheli P, Schwemmle M, Kochs G (2015) Mx GTPases: dynamin-like antiviral machines of innate immunity. Trends Microbiol 23(3):154–163
11. Xu B, Kong J, Wang X, Wei W, Xie W, Yu XF (2015) Structural insight into the assembly of human anti-HIV dynamin-like protein MxB/Mx2. Biochem Biophys Res Commun 456(1):197–201
12. Ramachandran R, Schmid SL (2018) The dynamin superfamily. Curr Biol 28(8): R411–R416
13. Ingerman E, Nunnari J (2005) A continuous, regenerative coupled GTPase assay for dynamin-related proteins. Methods Enzymol 404:611–619

Chapter 6

Purification of Farnesylated hGBP1 and Characterization of Its Polymerization and Membrane Binding

Linda Sistemich and Christian Herrmann

Abstract

The human guanylate-binding protein 1 (hGBP1) is the best characterized isoform of the seven human GBPs belonging to the superfamily of dynamin-like proteins (DLPs). As known for other DLPs, hGBP1 also exhibits antiviral and antimicrobial activity within the cell. hGBP 1, like hGBPs 2 and 5, carries a CAAX motive at the C-terminus leading to isoprenylation in the living cells. The attachment of a farnesyl anchor and its unique GTPase cycle provides hGBP1 the ability of a nucleotide- stimulated polymerization and membrane binding. In this chapter, we want to show how to prepare farnesylated hGBP1 ($hGBP1_{fn}$) by bacterial synthesis and by enzymatic modification, respectively, and how to purify the non-farnesylated, as well as the farnesylated hGBP1, by chromatographic procedures. Furthermore, we want to demonstrate how to investigate the special features of polymerization by a UV-absorption-based turbidity assay and the binding to artificial membranes by means of fluorescence energy transfer.

Key words Large GTPases, GBP, Farnesylation, Membrane binding, Polymerization, Enzymatic modification, Bacterial synthesis, FRET, Turbidity

1 Introduction

The human guanylate-binding proteins (hGBPs) represent large GTPases inducible by interferon II and exhibit antiviral and anti-bacterial effects [1]. The human guanylate-binding proteins (hGBPs) fulfill most of the criteria belonging to the dynamin superfamily, although they only share a small sequence homology with the other members [2]. The best characterized isoform of the seven human guanylate-binding proteins is hGBP1. It fulfills the minimum requirement of domain architecture by consisting of a large GTPase domain at the N-terminus, a middle domain, and a GTPase effector domain, with a total weight of around 68 kDa and a length of 592 amino acids. Like the other members of the superfamily, hGBP1 shows an oligomerization-dependent GTPase activation and a low GTP-binding affinity, rendering the help of guanine nucleotide-exchange factors and GTPase-activating

Rajesh Ramachandran (ed.), *Dynamin Superfamily GTPases: Methods and Protocols*, Methods in Molecular Biology, vol. 2159, https://doi.org/10.1007/978-1-0716-0676-6_6, © Springer Science+Business Media, LLC, part of Springer Nature 2020

proteins unnecessary [3–6]. In contrast to the other dynamin-related proteins, hGBP1 catalyzes the hydrolysis of GTP to GDP and GMP [7, 8]. Within the cell, hGBP1 carries a farnesyl anchor to interact with membranes to fulfill its purpose. But when synthesizing hGBP1 heterologously in *E. coli*, the lipid anchor is not attached, since bacteria do not carry out this posttranslational modification. Therefore, the enzyme farnesyltransferase (FTase) is needed. In order to obtain farnesylated hGBP1 (hGBP1$_{fn}$), the modification can occur by bacterial synthesis when cosynthesizing hGBP1 and FTase in *E. coli* [9]. Alternatively, an additional enzymatic modification can be carried out after the purification of nonmodified hGBP1. Farnesylated hGBP1 (hGBP1$_{fn}$) shows additional characteristics and abilities compared to the nonmodified protein (Fig. 1). In contrast to other isoprenylated proteins, the farnesyl anchor sticks tightly to the protein in the nucleotide unbound state. Only upon GTP binding is the farnesyl tail released from this intramolecular contact and becomes available for interactions. By now, hGBP1$_{fn}$ can go in two competing cycles. If membranes are present, hGBP1$_{fn}$ attaches to them via the farnesyl anchor and detaches after GTP is consumed. A longer-lasting membrane binding could be observed with the GTP-analogue GDP$*$AlF$_x$. In the absence of membranes, the released farnesyl

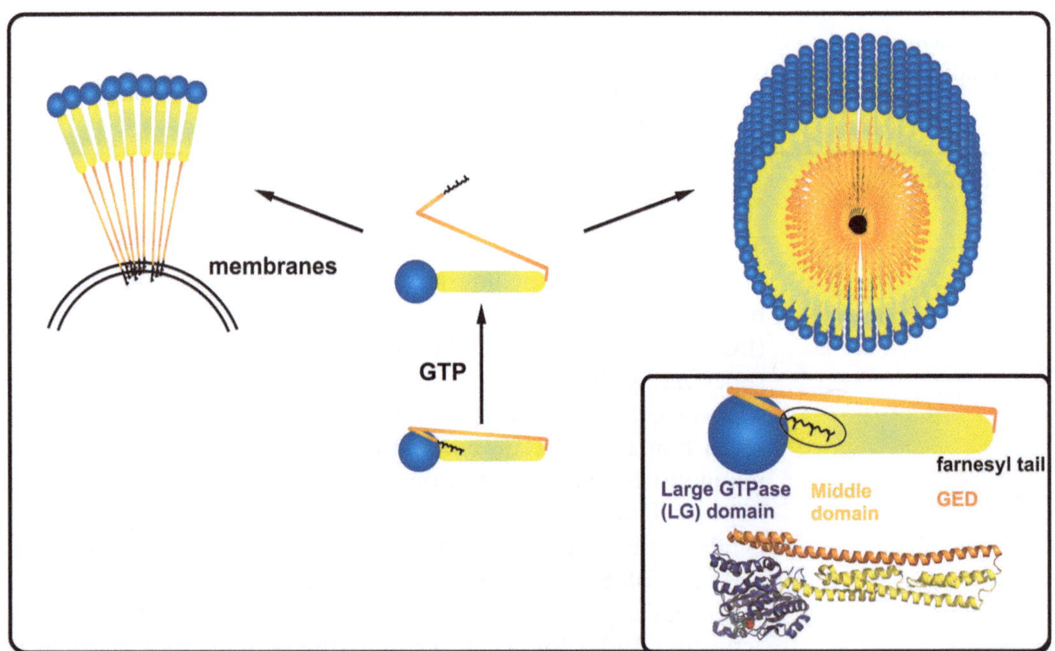

Fig. 1 Model of the special features of farnesylated hGBP1. Upon GTP binding, the farnesyl tail sticking to the protein is released, and the protein can fulfill two competing actions: The protein can bind to membranes (left side) or form a cylindrical polymer (right side). In the inset, the scheme is depicted according to the crystal structure (PDB:1F5N) with the addition of the farnesyl tail in black

anchor leads to reversible polymerization of $hGBP1_{fn}$, establishing an ordered supramolecular structure [10]. The phenomena of polymerization and membrane binding can be easily detected by a UV-absorption-based turbidity assay and fluorescence resonance energy transfer (FRET) measurements, respectively.

2 Materials

2.1 Biosynthesis of Non-Farnesylated and Farnesylated hGBP1 in E. coli

1. *E. coli* strain: BL21-CodonPlus (DE3)-RIL.
2. Vectors: pQE80L (ampicillin resistance) for hGBP1 and pRSFDuet-1 (kanamycin resistance) for FTase.
3. Terrific broth (TB) media: 12 g/l Bacto™ Tryptone, 24 g/l yeast extract, 4 ml/l glycerol, 0.017 M KH_2PO_4, and 0.072 M K_2HPO_4 (*see* **Notes 1** and **2**).
4. Lysogeny broth (LB) media: 10 g/l Bacto™ Tryptone, 5 g/l yeast extract, and 10 g/l NaCl, pH adjusted to 7.3–7.5 with NaOH (*see* **Note 2**).

2.2 Purification of Non-Farnesylated and Farnesylated hGBP 1

Prepare all buffer solutions with deionized water. After adjusting the pH value to 7.9 at 4 °C with HCl or NaOH, buffers have to be sterile filtered (0.22 μm), degassed, and stored at 4 °C afterward (*see* **Note 3**).

1. Lysis buffer: 50 mM Tris-HCl, 5 mM $MgCl_2$, 500 mM NaCl, and 10% (v/v) glycerol.
2. Buffer A: 50 mM Tris-HCl, 5 mM $MgCl_2$, and 500 mM NaCl.
3. Buffer B_{10}: 50 mM Tris-HCl, 5 mM $MgCl_2$, 150 mM NaCl, and 10 mM imidazole.
4. Buffer B_{150}: 50 mM Tris-HCl, 5 mM $MgCl_2$, 150 mM NaCl, and 150 mM imidazole.
5. Buffer C: 50 mM Tris-HCl, 5 mM $MgCl_2$, 150 mM NaCl, and 2 mM dithiothreitol (DTT).
6. Gua buffer (pH 6.5): 20 mM K_2HPO_4, 600 μM KH_2PO_4, and 6 M guanidinium hydrochloride.
7. High-salt buffer: 50 mM Tris-HCl, 5 mM $MgCl_2$, 1.2 M $(NH_4)_2SO_4$, and 2 mM DTT.
8. Low-salt buffer: 50 mM Tris-HCl, 5 mM $MgCl_2$, and 2 mM DTT.
9. Farnesylation buffer: 50 mM Tris-HCl, 5 mM $MgCl_2$, 150 mM NaCl.

2.3 Turbidity Assay

Buffer solutions used in experiments at 25 °C are prepared with deionized water and are adjusted to pH 7.9 at room temperature with HCl and NaOH before sterile filtration (0.22 μm) and degassing.

1. Buffer $C_{7.9}$: 50 mM Tris-HCl (pH 7.9), 5 mM $MgCl_2$, and 150 mM NaCl.
2. HPLC buffer (pH 6.4): 10 mM tetra-n-butylammonium bromide, 30 mM K_2HPO_4, 70 mM KH_2PO_4, and 0.2 mM sodium azide. Addition of 4% (v/v) acetonitrile after filtering (0.22 μm) and degassing (*see* **Note 4**).

2.4 LUV-Binding Assay

1. Buffer $C_{7.9,\ AlFx}$: 50 mM Tris-HCl, 5 mM $MgCl_2$, 150 mM NaCl, 300 μm $AlCl_3$, and 10 mM NaF (*see* **Note 5**).
2. Brain polar lipids (BPL) stock: 25 mg/ml in chloroform.
3. 1,2-Dipalmitoyl-sn-glycero-3-phosphoethanolamine-N-(lissamine rhodamine B sulfonyl) (ammonium salt) (Liss Rhod PE) stock: 1 mg/ml in chloroform.
4. Labelling buffer: 50 mM Tris-HCl (pH 7.4), 5 mM $MgCl_2$, and 150 mM NaCl.

3 Methods

3.1 Protein Biosynthesis

It is necessary to work under sterile conditions during the whole procedure of protein biosynthesis (*see* **Note 6**).

1. For biosynthesis the plasmid DNA (pQE80L with hGBP1 as insert) is transformed in BL21-CodonPlus (DE3)-RIL. For farnesylation by bacterial synthesis, BL21 are used, already containing the pRSFDuet-1 vector with FTase as insert (*see* **Note 7**). Transformation is performed via heat shock (*see* **Note 8**).
2. A single colony is taken to inoculate a 100 ml LB preculture, which is then incubated at 37 °C and 180 rpm for 14–16 h (*see* **Note 9**). The main culture is started by adding the preculture to TB media in a ratio of 1:80.
3. The main culture is incubated at 37 °C and 90 rpm until an optical density of 0.5–0.6 at 600 nm is reached. The temperature is reduced to 20 °C, and 100 μM isopropyl β-D-1-thiogalactopyranoside (IPTG) is added to initiate protein biosynthesis. The main culture was incubated for a further 14 h before harvesting the cells by centrifugation at 5000 × *g* for 15 min (*see* **Note 10**). The cell pellet can be used directly for purification or can be stored at −80 °C.

3.2 Purification

The first step of purification for both proteins (non-farnesylated and farnesylated) is the lysis of the cells. Therefore, the pellet is suspended in lysis buffer (10 ml lysis buffer per 1 g pellet) on ice, and the cells are mechanically broken by ultrasonification. Eight cycles of 1 min with a power of 30% sonification with pauses of 1–2 min are performed (*see* **Note 11**). To remove cell debris, the lysate is centrifuged for 45 min with 15,000 × *g* at 4 °C.

3.2.1 Nonmodified hGBP1

1. Since hGBP1 is expressed with an N-terminal His_6 tag, the first step of purification is an immobilized metal ion affinity chromatography (IMAC; tetradentate chelating agarose resin charged with divalent cobalt (Co^{2+})) (Fig. 2a, left panel). The supernatant of the lysate is applied to the column which is already equilibrated with buffer A. After washing with 5 CV buffer A, unspecifically bound proteins are eluted by washing with 2–3 CV buffer B_{10}. Elution of hGBP1 occurs over 2 CV of buffer B_{150}. All the steps are carried out with a flow rate of 2–2.5 ml/min (*see* **Note 12**). The eluted fractions are analyzed with SDS-PAGE (Fig. 3a), and the fractions containing the protein of interest are pooled and precipitated by the addition of ammonium sulfate (*see* **Note 13**).
2. A size-exclusion chromatography (SEC) is performed (Fig. 2c, left panel) (*see* **Note 14**), to gain the monomeric protein only and dispose of aggregates. The ammonium sulfate precipitate is dissolved in a minimum of buffer C and applied to the equilibrated column (1.5 CV, buffer C). Elution occurs over 1–2 CV with buffer C. Again, the eluted fractions are analyzed with SDS-PAGE (Fig. 3a) to identify proteins and purity. The fractions containing monomeric protein (*see* **Note 15**) are collected and concentrated to 0.5–1 mM by ultrafiltration with a cutoff of 10 kDa (*see* **Note 16**). The concentration is determined, and the protein is aliquoted, shock frozen, and stored at −80 °C (*see* **Note 17**).

3.2.2 Enzymatically Farnesylated hGBP1

The purification of farnesylated hGBP1 is based on the procedure described in [9].

1. In a 4 ml glass vial with teflon cap, 60 μM hGBP1 is incubated with 1.25 μM FTase and 150 μM farnesyl pyrophosphate (FPP) in farnesylation buffer for 14 h at 4 °C (*see* **Note 18**).
2. Hydrophobic interaction chromatography (HIC; butyl sepharose) is performed (Fig. 2b, middle panel) (*see* **Note 19**) to separate $hGBP1_{fn}$ from hGBP1 and FTase. After incubation, the reaction mixture is adjusted to a final concentration of 1.25 M $(NH_4)_2SO_4$ with 3 M $(NH_4)_2SO_4$ stock solution. The HIC column is equilibrated with 1.5 CV high-salt buffer, and the dissolved protein is applied to the column. To separate the proteins, the concentration of ammonium sulfate is decreased to 0 M (low-salt buffer) in the four successive steps. The concentration of high-salt buffer is initially reduced to 60% (increase to 40% low-salt buffer, respectively). After reaching a stable baseline, the concentration of the high-salt buffer is reduced to 45% in a continuous gradient over 3 CV. In the third step, the concentration of high-salt buffer is further decreased continuously to 25% over 3.75 CV. The last step is an

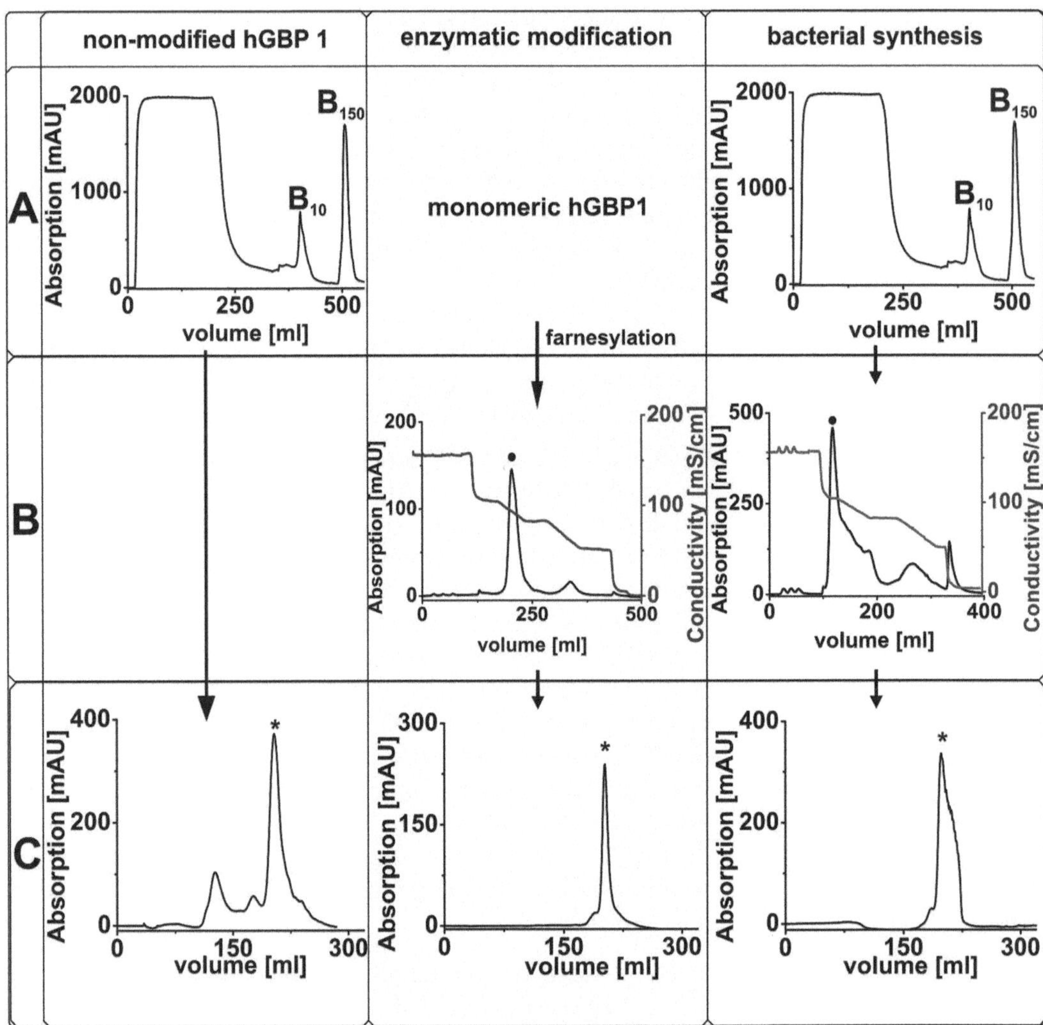

Fig. 2 Purification of farnesylated hGBP1 (bacterially synthesized and enzymatically modified) and non-modified hGBP1. *Left Panel*: Purification of non-modified protein. Co^{2+}-affinity chromatography to isolate His-tagged hGBP1 (**a**). Washing steps are performed with buffer A and buffer B_{10} (B_{10}), and elution is performed with buffer B_{150} (B_{150}). Isolation of monomeric protein (∗) occurs via SEC (**c**). *Middle panel*: Enzymatic farnesylation of hGBP1. Monomeric protein is farnesylated enzymatically (**a**). In the first step hGBP1$_{fn}$ (•) is purified by HIC (**b**). The stepwise salt gradient is reported by conductivity measurements of the eluted salt (blue). The gradient is performed by reducing salt concentration first to 60%, and then over 3 CV to 45% in the first gradient and over 3.75 CV to 25% in the second gradient. In the second step, farnesylated monomeric protein (∗) is isolated by SEC (**c**). *Right panel*: Bacterial synthesis of farnesylated hGBP1. Co^{2+}-affinity chromatography to isolate His-tagged hGBP1 (fn and non-fn) (**a**). Washing steps are performed with buffer A and buffer B_{10} (B_{10}), and elution is performed with buffer B_{150} (B_{150}). In the second step, hGBP1$_{fn}$ (•) is purified by HIC (**b**). The stepwise salt gradient is reported by conductivity measurements of the eluted salt (blue). The gradient is performed by reducing salt concentration first to 60% and then over 3 CV to 45% in the first gradient and over 3.75 CV to 25% in the second gradient. In the third step, farnesylated monomeric protein (∗) is isolated by SEC (**c**)

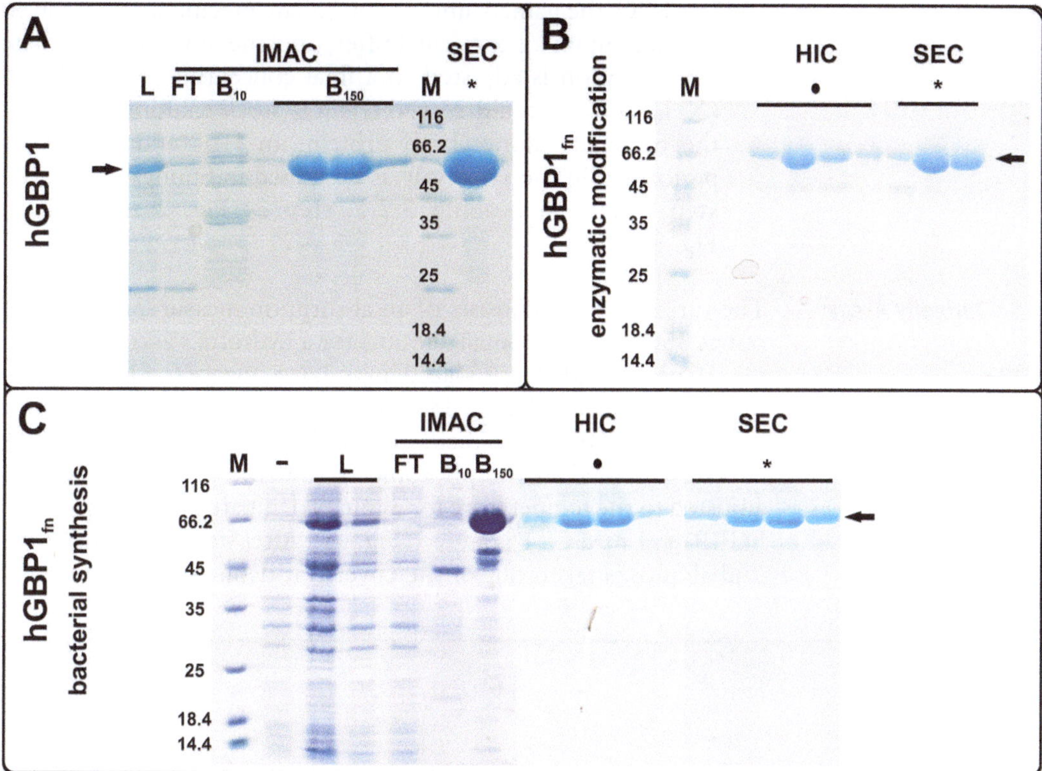

Fig. 3 SDS-PAGE documenting the purification of non-farnesylated and farnesylated hGBP1. The sizes of the depicted proteins are shown in kDa (M), and hGBP1 is indicated by an arrow. (**a**) Purification of non-farnesylated hGBP1 with IMAC and SEC. The lysate (L) is applied to the IMAC column, and the flow through (FT) is collected separately. Afterward, the column is washed with buffer B_{10} , and hGBP1 is eluted with buffer B_{150}. Monomeric hGBP1 is gained after SEC (∗). (**b**) Purification of enzymatically modified hGBP1$_{fn}$ with HIC and SEC. The farnesylated hGBP1 (●) is isolated first by HIC, and monomeric hGBP1$_{fn}$ (∗) is gained by SEC. (**c**) Purification of bacterially synthesized hGBP1$_{fn}$ with IMAC, HIC, and SEC. The overexpression of hGBP1 and FTase is shown in the lysate (L) compared to the lysate before addition of IPTG (IPTG). The lysate (L) is applied to the IMAC column, and the flow through (FT) is collected separately. Afterward, the column is washed with buffer B_{10}, and hGBP1 is eluted with buffer B_{150}. The farnesylated hGBP1 (●) is isolated first by HIC, and monomeric hGBP1$_{fn}$ (∗) is gained by SEC

immediate jump to 0% high-salt and 100% low-salt buffer. The eluted fractions are analyzed by SDS-PAGE (Fig. 3b), and the corresponding fractions with hGBP1$_{fn}$ are concentrated by ultrafiltration with a 10 kDa cutoff.

3. Size-exclusion chromatography is performed for the nonmodified protein (*see* Subheading 3.2.1, **step 2**) (Fig. 2c middle panel, Fig. 3b).

3.2.3 Bacterially Synthesized Farnesylated hGBP1

1. The first step corresponds to the IMAC of the non-modified protein (*see* Subheading 3.2.1, **step 1**) (Fig. 3c).

2. For HIC, the gained ammonium sulfate precipitate is dissolved in a minimum of low-salt buffer, and the ammonium sulfate concentration is adjusted to a final concentration of 1.25 M $(NH_4)_2SO_4$ with a 3 M $(NH_4)_2SO_4$ stock solution. Also, for the bacterially farnesylated protein, an HIC and an SEC is performed in the same way as described in Subheading 3.2.2, **step 2** and Subheading 3.2.1, **step 2** (Fig. 2 right panel, Fig. 3c).

3.3 Turbidity Assay

The turbidity assay consists of an absorption measurement carried out at 350 nm, while small aliquots for a hydrolysis assay are taken.

10 μM of hGBP1$_{fn}$ is placed in a quartz cuvette (*see* **Note 20**) in buffer $C_{7.9}$ with 50 μM BSA (*see* **Note 21**). The total volume amounts to 150 μl. The measurement is carried out at 25 °C. The cuvette is placed in the photometer, and the protein solution is adjusted to the temperature for 5 min. To initiate polymerization of hGBP1$_{fn}$, 1 mM GTP is added (Fig. 4a). After starting the reaction, 2 μl aliquot is taken out of the cuvette at defined time points and

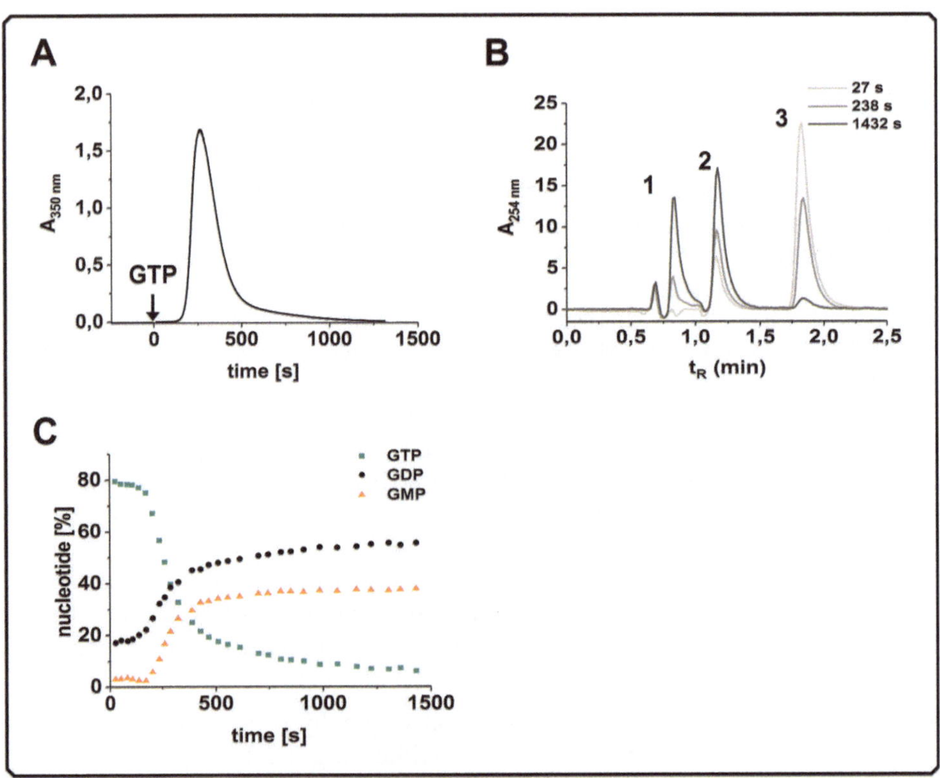

Fig. 4 Turbidity assay. (**a**) 10 μM hGBP1$_{fn}$ is incubated for 5 min in buffer $C_{7.9}$ with 50 μM BSA at 25 °C before initiating polymerization by adding 1 mM GTP. The absorption is measured at 350 nm. (**b**) Samples taken during the absorption measurement are analyzed by HPLC (RP-18e, 10–4.6 mm) in HPLC buffer. GMP (1), GDP (2), and GTP (3) are separated and detected at 254 nm. (**c**) Time trace of GTP hydrolysis while polymerization. HPLC analysis (**b**) yields the quantities of GMP, GDP, and GTP

mixed in 30 μl HPLC buffer. The samples are further heated for 5 min at 80 °C. After that, samples can be further analyzed by HPLC or stored at −20 °C.

The samples taken from the absorption cuvette and diluted with HPLC buffer are spun down and applied to HPLC column (RP-18e, 10–4.6 mm) (*see* **Note 22**) to analyze the composition of the nucleotides (Fig. 4b). HPLC buffer is used to operate the column with a flow rate of 4.5 ml/min to separate GTP, GDP, and GMP. The elution of the nucleotides is detected at 254 nm (Fig. 4b). The integration of the elution profile yields the concentration of the three nucleotides. Altogether, the time course of GTP hydrolysis and of GDP and GMP formation is obtained, respectively (Fig. 4c).

3.4 LUV-Binding Assay

The binding of farnesylated hGBP1 to artificial membranes can be tracked by using a FRET-based method. Therefore, lipids carrying a rhodamine label are used, as well as a mutant of hGBP1$_{fn}$ (hGBP1-Q577C$_{fn}$) labelled with Alexa Fluor 488 dye (*see* **Note 23**).

3.4.1 LUV Preparation

Large unilamellar vesicles (LUVs) have to be prepared freshly every day to ensure that they are intact and not oxidized in the buffer.

1. To prepare a stock solution of 2 mg/ml LUVs with a labelling efficiency of 0.5% rhodamine-labelled lipids, 80 μl BPL and 10 μl Liss Rhod PE are combined in a glass tube (*see* **Note 24**) and dried with a slow flow of argon gas while rotating the tube to create a homogenous thin lipid film on the bottom of the tube (*see* **Note 25**).
2. The tube is placed in vacuum for 30 min to remove chloroform completely.
3. The lipid film is covered with 1 ml buffer $C_{7.9}$; the tube is sealed with parafilm and placed in a water bath at 55 °C for 30 min. The solution is vortexed every 10 min to improve hydration. After the incubation, the lipid film is restlessly resuspended with a pipette and transferred into a 1.5 ml reaction tube. Three to five freeze and thaw cycles are performed in liquid nitrogen and a water bath at 55 °C, respectively, to ensure proper vesicle formation. To prevent aggregate formation, the solution is vortexed after each thawing step. The LUVs of a defined size are yielded by taking up the lipid solution in 1 ml glass syringes and performing 10 extrusion cycles (*see* **Note 26**) at 50 °C with a membrane of a defined pore size (100 nm) (*see* **Note 27**).

3.4.2 Protein Labelling

Labelling of hGBP1-Q577C_{fn} is performed in labelling buffer (pH 7.4 adjusted at 4 °C). Attachment of the fluorophore to the protein is maintained by a maleimide coupling with an available cysteine of hGBP1-Q577C_{fn}.

1. Dithiothreitol (DTT) is removed from protein solution by performing a buffer exchange from buffer C to labelling buffer by ultrafiltration (*see* **Note 28**).
2. The concentration is determined (*see* **Note 28**), and the protein is incubated with an equimolar amount of Alexa Fluor 488 dye on ice for 10 min to the exclusion of light. The reaction is then stopped by adding 2 mM DTT, and unreacted dye is removed via ultrafiltration (*see* **Note 29**). Again, the labelled protein is concentrated up to 100–200 μM, usually in a final volume of 100–200 μl.
3. Final concentration of the protein and labelling efficiency (LE) of the fluorophore is determined by measuring the absorption at 280 nm and 491 nm (ε_{491}: 71,000 1/(mol∗cm) and ε_{280}: 45,400 1/(mol∗cm), respectively) and setting them into relation according to formula 1 (*see* **Note 30**). The labelled protein (hGBP1-Q577C_{fn}∗488) can also be shock frozen in liquid nitrogen and stored at −80 °C.

$$\mathrm{LE} = \frac{\frac{A_{491}}{\varepsilon_{491}}}{\frac{A_{280} - A_{491} \times 0.11}{\varepsilon_{280}}} \tag{1}$$

3.4.3 Measurement

In general, the measurements (Fig. 5) are performed in a special quartz cuvette for fluorescence measurements. The total volume

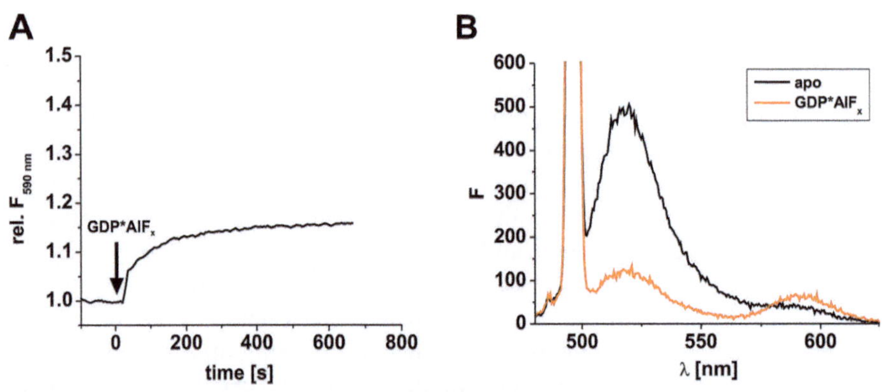

Fig. 5 LUV binding assay. (**a**) 2.5 μM hGBP1-Q577C_{fn}∗488 is incubated with 0.5 mg/ml LUVs (*d*: 100 nm) and 50 μM BSA in buffer $C_{7.9,\ AlFx}$ for 5 min before adding 250 μM GDP∗AlF_x. Measurements are performed at 25 °C by exciting at 495 nm and detecting at 590 nm. Both slits are set to 2.5 nm. Change in fluorescence is displayed relative to initial fluorescence. (**b**) An emission spectrum starting at 480 nm up to 650 nm is performed; exciting at 495 nm is taken to observe the changes in donor and acceptor emission in the presence and absence of GDP∗AlF_x

amounts to 150 µl. For the time drive measurement (Fig. 5a), the excitation wavelength is set to 495 nm, while the emission wavelength is set to 590 nm. Both slits are opened to 2.5 nm.

0.5 mg/ml LUVs with 2.5 µM hGBP1-Q577C_{fn}*488 in buffer $C_{7.9,\ AlFx}$ with 50 µM BSA is incubated in a quartz cuvette for 5 min at 25 °C (*see* **Notes 21** and **31**). Binding is induced by adding 250 µM of GDP*AlF_x (*see* **Note 32**) and mixing properly.

Binding of the protein to the membrane will lead to an increase in the detected acceptor fluorescence (Fig. 5a) and the decrease in the donor fluorescence due to the resonance energy transfer (Fig. 5b).

Prior and after addition of GDP*AlF_x, a scan of the emission spectra of Alexa Fluor 488 dye and rhodamine is performed. The range of 480 nm to 650 nm is scanned with a speed of 200 nm/min while exciting the sample at 495 nm. Both slits are opened to 2.5 nm (Fig. 5b).

4 Notes

1. While preparing TB media, it is important to autoclave the solution containing yeast extract and Bacto™ Tryptone and glycerol separated from the salt solution to avoid precipitation. Therefore 4.5 l of yeast extract, Bacto™ Tryptone, and glycerol and 0.5 l of phosphate salts (containing 0.17 M KH_2PO_4 and 0.72 M K_2HPO_4) are prepared separately and mixed after autoclaving.

2. Before usage, the media is enhanced with either 100 µg/l ampicillin, when only using vector pQE80L, or 50 µg/l ampicillin plus 12.5 µg/l kanamycin, when using both vectors pQE80L and pRSFDuet-1.

3. The pH value has to be adjusted at the temperature of 4 °C because the pH of Tris-buffers is temperature dependent. Therefore, it is sufficient to precool double-distilled water (ddH_2O). Salt solutions can be added at room temperature. Dithiothreitol (DTT) has to be added the same day as usage.

4. The pH value should be perfectly adjusted, when weighing in all the components very carefully. The accuracy of the pH value is very important since the separation of the nucleotides is highly dependent on the pH value. For 10 l of buffer, the following masses are needed: 32.2 g tetra-n-butylammonium bromide, 57.48 g K_2HPO_4, 93.25 g KH_2PO_4, and 0.13 g sodium azide. Acetonitrile (4%) are added directly before usage.

5. To prepare buffer $C_{7.9,\ AlFx}$, 300 µM $AlCl_3$ and 10 mM NaF have to be added freshly the same day to buffer $C_{7.9}$.

Thereafter, buffer $C_{7.9,\ AlFx}$ needs to be sterile filtered again. Next to the buffer preparation, the stock solutions of the respective salts (500 mM NaF and 300 mM $AlCl_3$) have to be prepared the same day also. Any predilution of the protein or nucleotide stock should only take place in buffer $C_{7.9}$. Buffer $C_{7.9,\ AlFx}$ is exclusively the measuring buffer.

6. To ensure sterile conditions, the chimney effect of a flame is used. Therefore, the ground on which the Bunsen burner is placed is properly cleaned, and the work is performed close to the flame.
7. *E. coli* were transformed via heat shock with a pRSFDuet-1 vector with FTase as an insert. A single colony was taken and used to create new competent cells with the $CaCl_2$ method [11].
8. Plasmid DNA is added to the cells (100 μl) and incubated on ice for 30 min. Afterward, a heat shock for 40–50 s at 42 °C is performed. The cells are placed back on ice and 400 μl of LB media without antibiotics are added. The cells are incubated for 1 h at 37 °C, before pelleting them, discarding ca. 400–450 μl of supernatant and resuspending the cells. The resuspended cells are plated to grow single colonies.
9. An incubation time of 14–16 h refers to an overnight incubation.
10. Overexpression is confirmed with SDS-PAGE. Therefore, samples are taken before addition of IPTG and before harvesting the cells (after incubation of 14 h).
11. While disrupting the cell walls via sonification, it is important to place pauses in between the sonification cycles. The lysate is heated up during the process, and it is important to keep the temperature below 8 °C to avoid denaturation. Therefore, it is best to add some NaCl to the ice bath.
12. The flow rate has to be adjusted to the pressure limit of the respective column. The slower the flow rate is adjusted for eluting the protein (down to 0.5 ml/min), the sharper the elution peak gets.
13. For ammonium sulfate precipitation, the collected fractions are combined and 3 M ammonium sulfate is added in small portions, while stirring on ice. The solution is transferred to a 50 ml reaction tube, and the precipitation is pelleted at 8000 × g at 4 °C for 10 min. The supernatant is discarded, and the remaining buffer is removed from the tube walls with a wipe. Ammonium sulfate pellets are stored at −80 °C.
14. The type of SEC column material has to be chosen corresponding to the size of protein. Superdex® 200 26/60 was used for purification.

15. Monomeric size is defined by a calibration of the column with a set of reference proteins.
16. It is said that using a cutoff filter half the size of your protein will ensure that it will not pass the membrane. This is true for allover globular proteins. The human guanylate-binding protein 1 (hGBP1) has a size of 68 kDa, but since it has an elongated shape, a cutoff filter of 30 kDa is not suitable for concentrating hGBP1. Therefore, a cutoff of 10 kDa is used.
17. Determination of protein concentration is performed in Gua buffer. Usually the protein is diluted to roughly 200 μM to get a proper signal (ε_{280} (hGBP1, Gua buffer): 43,240 l/(mol∗cm)).
18. An excess of FPP is needed to gain a maximum yield of farnesylated protein. Furthermore, the concentration of FPP and, therefore, of hGBP1 cannot be increased since FPP is solved in methanol partly. A concentration of 5% methanol, in total, in the reaction mixture should not be exceeded to ensure stability of the proteins. In addition to that, it is important to carefully add FPP as the very last component to the reaction mixture to ensure low concentration of methanol.
19. FTase is also expressed with an N-terminal His_6 tag.
20. It is preferable to use glass rather than polystyrene cuvettes to minimize sticking of the protein to the walls.
21. BSA is added to comfort the protein in small concentrations and to avoid sticking to the walls.
22. A guard cartridge (RP-18e, 10–4.6 mm) is used to hold back protein from the HPLC column.
23. An additional cysteine at the position 577 is introduced to guarantee a labelling position close to the farnesyl anchor and, therefore, close to the membrane.
24. A special glass syringe is used to transfer the lipids. If only one syringe is provided, make sure to transfer BPL first followed with rhodamine-labelled lipids to avoid fluorescence contamination in the respective stock solution.
25. If no argon gas is available, compressed air is also fine. The lipid film should only cover an area of the glass tube, small enough that it will be completely covered with buffer $C_{7.9}$ in the hydration process.
26. Performing one cycle implies passing the membrane twice to have the solution back in the uptaking syringe.
27. To avoid the disruption of the fragile membrane, it is important to protect it from both sides with a membrane support. While assembling the extruder, the membrane and membrane supports have to be moistened with buffer $C_{7.9}$ to make it stick

together. Additionally, it is helpful to flush the extruder and the syringes with buffer $C_{7.9}$ first to hydrate the membrane and the membrane supports adequately and to minimize the loss of lipids while extruding. After performing the extrusion cycles, it is important to take the lipids from the syringe, which was not used for taking up the lipids from the reaction tube, to make sure that there are no impurities from the very first extrusion process in your final sample. In contrast to the temperature of the water bath at 55 °C, by now, the temperature is reduced to 50 °C to ensure a longer lifetime of the glass syringes, being sensitive to temperature >50 °C.

28. Usually the amount of protein used for labelling is chosen to yield a final volume of 100–200 μl with 200–100 μM of protein in order to have a sufficient amount of protein for a test series and to have an easy concentration determination without having to dilute the protein (ε_{280} (hGBP1, buffer C): 45,400 l/(mol∗cm)). This time, the determination of concentration is referenced to buffer $C_{7.9}$. A proper removal of DTT is reached by mixing 20–30 μl of concentrated protein with 4 ml of labelling buffer. Therefore, labelling buffer is placed first in the ultrafiltration tube to which the protein is added.
29. Buffer C (4 ml) containing 2 mM DTT is placed in the ultrafiltration tube, and the reaction mixture is added to inactivate still reactive fluorophore. The same ultrafiltration tube used for the buffer exchange can be used when flushed properly with buffer C afterward.
30. Since the fluorophore is also absorbing gradually at 280 nm, the fluorophore- specific correction factor (0.11) has to be taken into account. Volume and concentration are again chosen to determine concentration without any dilution (referenced to buffer $C_{7.9}$).
31. Fluorescence of the dyes can be very temperature sensitive. To ensure a constant baseline in the beginning, it is important to equilibrate the temperature to 25 °C of the sample to be measured, as the protein was previously stored on ice.
32. The formation of the GTP transition-state analogue GDP∗AlF_x occurs immediately by the addition of 250 μM GDP to buffer $C_{7.9,\ AlFx}$.

References

1. Praefcke GJK (2017) Regulation of innate immune functions by guanylate-binding proteins. Int J Med Microbiol 308(1):237–245
2. Praefcke GJK, McMahon HT (2004) The dynamin superfamily: universal membrane tabulation and fission molecules? Nat Rev Mol Cell Biol 5(2):133–147
3. Praefcke GJK et al (1999) Nucleotide-binding characteristics of human guanylate-binding protein 1 (hGBP1) and identification of the

third GTP-binding motif. J Mol Biol 292 (2):321–332

4. Ghosh A et al (2006) How guanylate-binding proteins achieve assembly-stimulated processive cleavage of GTP to GMP. Nature 440 (7080):101–104
5. Prakash B et al (2000) Structure of human guanylate-binding protein 1 representing a unique class of GTP-binding proteins. Nature 403(6769):567–571
6. Kunzelmann S et al (2005) Nucleotide binding and self-stimulated GTPase activity of human guanylate-binding protein 1 (hGBP1). In: Balch WE, Der CJ, Hall A (eds) GTPases regulating membrane dynamics. Methods in enzymology, vol 404. Elsevier Academic Press, London, pp 512–527
7. Schwemmle PSM (1994) The interferon-induced 67-kDa guanylate-binding protein (hGBP1) is a GTPase that converts GTP to GMP. J Biol Chem 268:11299–11305
8. Kunzelmann S et al (2006) Transient kinetic investigation of GTP hydrolysis catalyzed by interferon-gamma-induced hGBP1 (human guanylate binding protein 1). J Biol Chem 281(39):28627–28635
9. Fres J et al (2010) Purification of the CaaX-modified, dynamin-related large GTPase hGBP1 by coexpression with farnesyltransferase. J Lipid Res 51(8):2454–2459
10. Shydlovskyi S et al (2017) Nucleotide-dependent farnesyl switch orchestrates polymerization and membrane binding of human guanylate-binding protein 1. Proc Natl Acad Sci 114(28):5559–5568
11. Cohen SN et al (1972) Nonchromosomal Antibiotic Resistance in Bacteria: Genetic Transformation of Escherichia coli by R-Factor DNA. Proc Natl Acad Sci 69 (8):2110–2114

Part III

Biophysical and Cellular Characterization of DSPs

Chapter 7

Microscale Thermophoresis (MST) as a Tool for Measuring Dynamin Superfamily Protein (DSP)–Lipid Interactions

Nikhil Bharambe and Rajesh Ramachandran

Abstract

Microscale thermophoresis (MST) is a robust new fluorescence-based technology that enables measurement of biomolecular interactions and binding affinities (K_D). MST is an immobilization-free alternative to surface plasmon resonance (SPR) and is cost-effective relative to isothermal titration calorimetry (ITC). In this chapter, using Drp1 as an example, we demonstrate for the first time, the application of MST to the determination of DSP–lipid interactions and the accurate measurement of K_D under physiologically relevant solution conditions.

Key words Dynamin-related protein 1, Drp1 variable domain, Microscale thermophoresis, Equilibrium dissociation constant, Protein–lipid interactions

1 Introduction

Microscale thermophoresis (MST) detects and quantifies biomolecular interactions based on the directed motion of fluorescent molecules (in this chapter, fluorescently labeled proteins) induced by a microscopic temperature gradient [1]. This thermophoretic effect is extremely sensitive to changes in the molecule's hydration shell, charge, size, or conformation influenced by intermolecular interactions [2]. Similar to ITC, MST is a titration-based approach that enables measurement of affinity constants in the binding equilibrium of a wide variety of interactions, including protein–lipid association. Unlike ITC, however, MST does not require very high concentrations of sample material and is not limited by buffer conditions or interactions that produce negligible changes in enthalpy not measurable by ITC. Moreover, unlike SPR, which relies on surface immobilization of biomolecules that could potentially interfere with binding, MST is performed under solution equilibrium conditions with free access to all interaction surfaces. Relative to ITC and SPR, MST is also very quick (a complete binding isotherm can be obtained in a matter of 10 min) and can

Rajesh Ramachandran (ed.), *Dynamin Superfamily GTPases: Methods and Protocols*, Methods in Molecular Biology, vol. 2159, https://doi.org/10.1007/978-1-0716-0676-6_7,

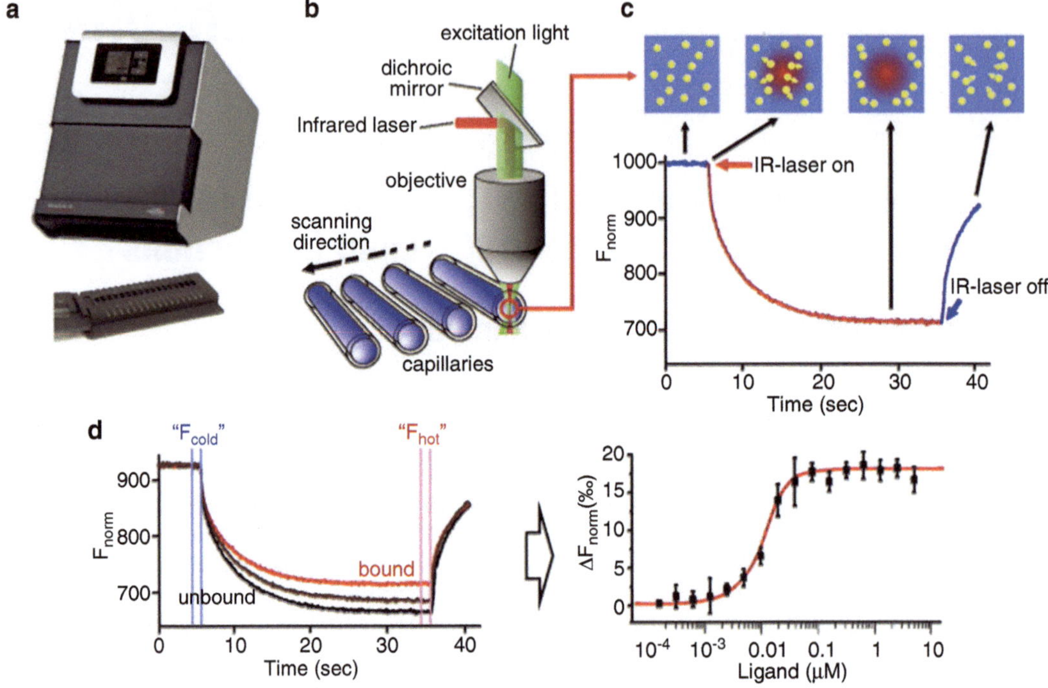

Fig. 1 (**a**) NanoTemper's Monolith™ NT.115 MST instrument with a capillary tray holding 16 capillaries shown below. (**b**) Schematic representation of MST optics. (**c**) Typical signal of an MST experiment. (**d**) MST binding experiment. Time traces (panel **c**) are analyzed, and thermophoresis is plotted as change in normalized fluorescence versus ligand concentration (panel **d**). Image adapted from Jerabek-Willemsen et al. Journal of Molecular Structure 1077 (2014) 101–113 for demonstrative purposes

reliably determine K_D (binding affinity) in the pM to mM range compared to ITC (nM to sub mM) and SPR (sub nM to low mM). Further, the high sensitivity of the MST approach coupled to low-volume fluidics necessitates only microliters of material at nanomolar concentrations.

The MST experimental setup is based on the Monolith™ NT.115 instrument (NanoTemper Technologies GmbH) (Fig. 1a). It comprises a system of 16 glass capillaries containing the same concentration of the fluorescently labeled molecule (e.g., labeled protein) and varying concentrations (achieved by serial dilution) of the unlabeled binding partner (e.g., lipid) (Fig. 1b). The MST instrument detects the motion of fluorescently labeled molecules along a microscopic temperature gradient (typically a temperature increment of 2–6 K on a length scale of 200 μm) induced by an infrared (IR) laser focused into the sample (Fig. 1b). The fluorescence excitation/emission optical path is aligned with the IR beam and allows for a very precise monitoring of the thermophoresis-dependent depletion of fluorophores along this temperature gradient (Fig. 1b, c). Emission is collected through the same objective used for the IR laser (Fig. 1c). To determine binding affinity, 16 capillaries containing a constant concentration of the fluorescently labeled molecule (in our case, labeled Drp1 or the Drp1 variable domain (VD)) and increasing concentrations of the unlabeled binding partner (in our case, liposomes) are scanned consecutively for thermophoresis (Fig. 1b, d). Changes in the thermophoresis of the fluorescent (protein) molecules upon ligand (lipid) binding are measured as a function of ligand concentration and are used to calculate K_D (Fig. 1d). In this chapter, using MST, we demonstrate determination of the binding affinity (or K_D) of full-length human Drp1 (or Drp1 VD) for target cardiolipin (CL) in liposomes [3]. The experimental scheme outlined below can be adapted for other DSPs to determine their respective lipid specificities and relative binding affinities.

2 Materials

2.1 Microscale Thermophoresis

1. NanoTemper's Monolith™ NT.115 instrument.
2. Monolith™ NT.115 Series Premium Capillaries ($n = 50$).
3. HEPES-buffered saline (HBS): 20 mM HEPES at pH 7.5 and 150 mM KCl. Prepare 800 μL.
4. MST buffer: HBS with 2 mg/mL BSA. Prepare 800 μL.
5. Surface Cys-modified, BODIPY™ FL- (BODIPY™ FL C_1-Iodoacetamide; ThermoFisher Scientific Catalog No. D6003)-labeled DSP prepared in HBS according to established procedures [3–5]. MST is demonstrated in this chapter using BODIPY™ FL-labeled full-length Drp1 and Drp1 VD (~2 mol of dye per mol of protein) that retain activity of the corresponding unlabeled molecules. Other site-directed dyes depending on the excitation capabilities of instrument may be used as long as dye labeling does not interfere with either protein structure or function (*see* **Note 1**). A 5 μM or greater stock concentration of fluorescently labeled protein in a volume of 200 μL is desired.
6. Freshly prepared liposomes or lipid nanotubes of a defined composition were prepared by either extrusion or sonication, as detailed elsewhere [3, 4, 6–8]. For the purposes of demonstration in this study, bovine heart cardiolipin (mostly 18:2 tetralineoyl-CL (TLCL)) that specifically binds both Drp1 and Drp1 VD [3, 5] and dioleoylphosphatidylcholine (DOPC) that serves as negative control, both purchased from Avanti Polar Lipids (Alabaster, AL), were used. Briefly, a 4 mM (final total lipid) stock of liposomes containing either 100% CL or 100% DOPC was prepared in a final volume of 100 μL by first drying an appropriate volume (amount) of each lipid from a chloroform stock to a thin film in a glass test tube under a

stream of nitrogen gas. Residual chloroform was removed under vacuum and heat (37 °C) in a SpeedVac for ~1–2 h. The dried lipid film was rehydrated in 100 μL of HBS for 30 min at 37 °C (in a water bath with gentle vortexing of the test tube every 5 min). The lipid suspension was subjected to three freeze–thaw cycles (using liquid nitrogen) and sonicated for 2 min in an Avanti® sonicator water bath (Avanti Polar Lipids, Alabaster, AL). Liposomes were briefly stored on ice until use (*see* **Note 2**).

7. 50 PCR tubes (0.2 mL capacity).

3 Methods

Carry out all procedures at room temperature unless otherwise specified.

3.1 Microscale Thermophoresis

1. Transfer 15 μL of MST buffer to each of the 16 PCR tubes (labeled 1–16) to perform serial dilution.
2. Add 15 μL of freshly prepared liposomes from the 4 mM liposome stock to tube 1 and mix thoroughly by gentle pipetting to achieve twofold dilution.
3. Likewise, transfer 15 μL from tube 1 to tube 2 and mix thoroughly to achieve fourfold dilution. Repeat process in sequence for tubes 3–16 to achieve serial dilution. Discard 15 μL of aspirated solution from the 16th tube. This will generate a lipid concentration range between 2 mM (tube 1) and 61 nM (tube 16) in twofold decrements (*see* **Note 3**).
4. Next, add 15 μL (equal volume) of either 0.2 μM BODIPY™ FL-labeled full-length Drp1 or 1.8 μM BODIPY™ FL-labeled Drp1 VD, both diluted in HBS from stock to each tube and mix thoroughly by gentle pipetting.
5. Cap the tubes, cover with aluminum foil, and incubate for 30 min at room temperature (*see* **Notes 4–6**).
6. After incubation, load each sample into Monolith™ NT.115 Premium Treated Capillaries (NanoTemper Technologies) by immersing one capillary per tube (*see* **Notes 7** and **8**).
7. Once the capillaries are filled, transfer them to the capillary cassette tray (cassette) (Fig. 1a).
8. Measure MST using the Monolith™ NT.115 instrument (NanoTemper Technologies) at 25 °C using the MST data acquisition software according to manufacturer's instructions. Choose the correct LED excitation source (blue LED for BODIPY™ FL), and adjust instrument parameters (in our case, 20% LED power and high MST (IR) power) (*see* **Notes 9** and **10**).

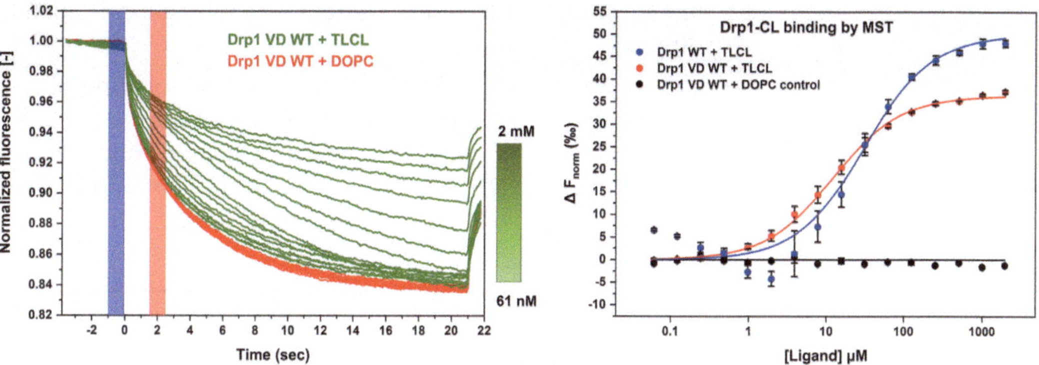

Fig. 2 Drp1–CL binding affinity revealed by MST. (**a**) Representative MST time traces for BODIPY™ FL-labeled Drp1 VD WT binding to TLCL (green traces) versus control DOPC liposomes (red traces) as a function of increasing lipid concentration. Blue- and red-shaded vertical columns represent F_{cold} and F_{hot} regions, respectively, used to calculate F_{norm}. (**b**) Dose–response curves for TLCL binding of BODIPY™ FL-labeled full-length Drp1 WT (0.1 μM final) and Drp1 VD WT (0.9 μM final). Data points represent the mean of three independent experiments. ΔF_{norm} (‰) is plotted versus lipid concentration. Error bars indicate SEM ($n = 3$)

3.2 Data Analysis and Binding Affinity Measurement

1. Immediately after starting the measurement, on-the-fly data analysis can be performed using the MST-analysis software according to the manufacturer's instructions. *See* manual at https://nanotempertech.com/monolith/.
2. For each capillary (each measurement point), an MST trace is recorded. All traces are then normalized to start at 1. A normalized trace is shown in Fig. 2a. For each trace, the F_{norm} value for the dose–response curve is calculated by dividing F_{hot} by F_{cold} (shaded regions in Fig. 2a). F_{hot} is the fluorescence value measured in the heated state, while F_{cold} is the fluorescence value measured in the cold state before the IR laser is turned on. Both values are averages taken between two time points in F_{hot} and F_{cold} regions. F_{norm} values are plotted as parts per thousand (‰). The change in normalized fluorescence (ΔF_{norm} in [‰]) is calculated by subtracting the F_{norm} baseline value from all data points of the same curve. In order to calculate the fraction bound, all ΔF_{norm} values of a curve are divided by the curve amplitude, resulting in the fraction bound (from 0 to 1) for each data point. For binding affinity analysis, ligand-dependent changes in MST are plotted as either fraction bound or ΔF_{norm} values (‰) versus ligand concentration in a dose–response curve.
3. Repeat **steps 1–8** for more replicates. Data from at least three independent measurements should be analyzed using the signal from *MST-off* time of −1 to 0 s and *MST-on* time of 1.5–2.5 s.
4. Data can be analyzed either using MO.Affinity Analysis software version 2.2.7 (NanoTemper Technologies) or by fitting

the data externally using Origin® graphing and analysis software.

5. A K_D of 36.3 ± 14.4 μM was obtained for the binding of full-length Drp1 (0.1 μM final) to 100% CL liposomes versus a K_D of 12.7 ± 4.5 μM for Drp1 VD at a higher concentration (0.9 μM final) (Fig. 2b). No binding was observed for Drp1 VD with control 100% DOPC liposomes.

4 Notes

1. Ensure that neither structure nor function of DSP is affected by dye labeling using enzymatic, self-assembly, and complementary membrane binding assay. Perform in comparison to unlabeled molecule.
2. Sonicated liposomes, which are generally <50 nm in diameter, are under high curvature stress and prone to fusion. Therefore, they must be used immediately after production.
3. The lipid concentration range and serial dilution can be adjusted according to the expected K_D.
4. Fluorescent dyes are highly sensitive even to ambient light. Labeled samples should be protected from photo damage by keeping in dark.
5. Never prepare small volumes (e.g., 20 μL) in large microreaction tubes (e.g., 500 μL or more). The high surface to volume ratio can lead to sample loss due to surface adsorption of proteins. Always use the smallest microreaction tubes possible (e.g., PCR tubes) or low-volume microwell plates.
6. Capillaries only need ~4 μL of sample, but do not prepare less than 20 μL of sample for each condition to ensure sample availability in case of accidental capillary breakage or drops.
7. Solution enters the capillary by capillary force. Therefore, avoid bubbles in the sample. Also, do not touch any part of the capillary with bare, greasy fingers as that may affect fluorescence detection.
8. Monolith™ NT.115 Series Premium Capillaries are preferred over standard capillaries for DSPs. Standard capillaries show adsorption of fluorescently labeled Drp1 onto the capillary wall (Fig. 3). To determine which capillary is best suited for your needs, take a closer look at the initial capillary scan. If the maxima of all the scanned capillaries perfectly overlay and are shaped like an inverse "U" (as seen for premium capillaries in Fig. 3), the capillaries are well suited. If the profiles show deviation among capillaries and are "M" shaped (split or

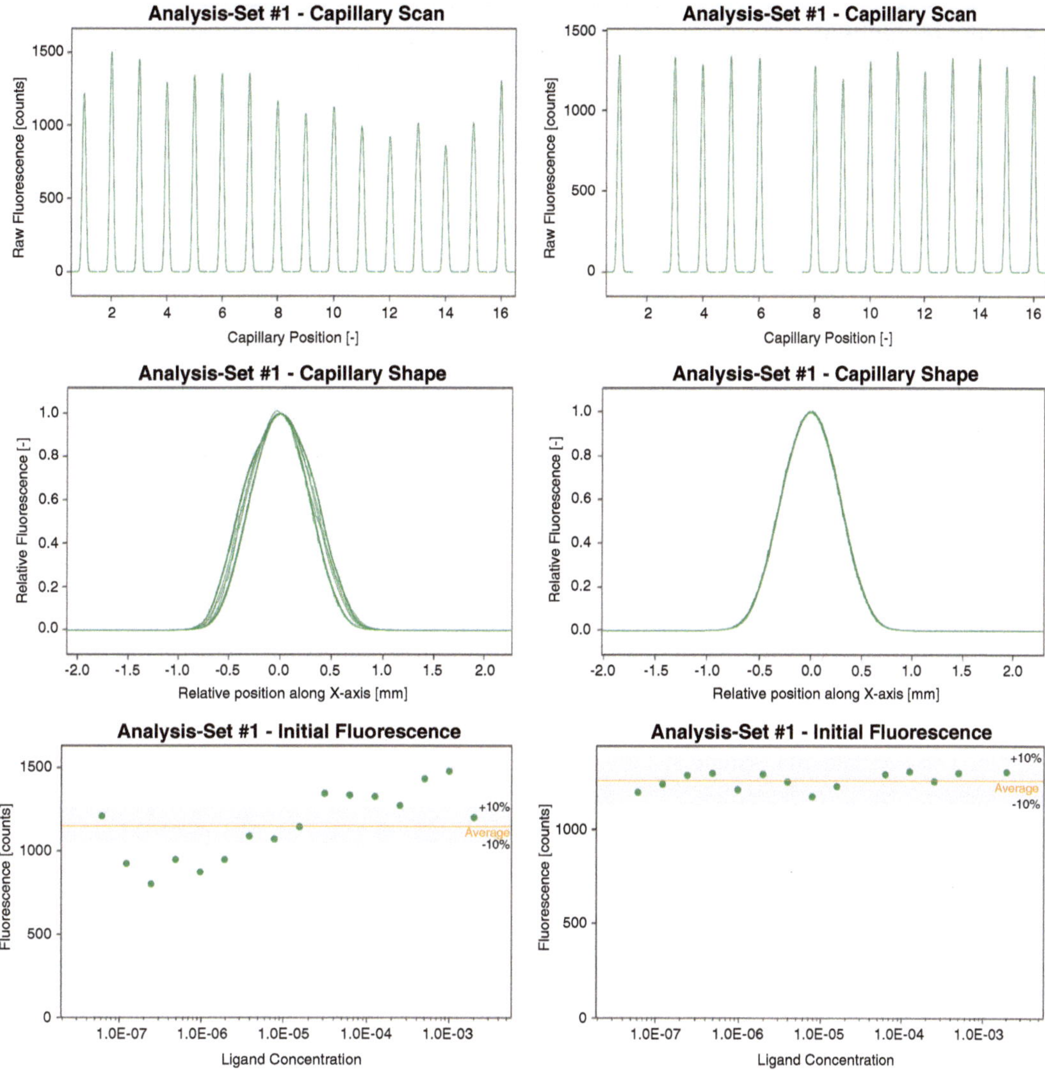

Fig. 3 Capillary scan (top), overlaid intensity shape profile (middle), and initial fluorescence intensity (bottom) of standard Monolith™ NT.115 Capillaries (left panels) versus Monolith™ NT.115 Premium Capillaries (right panels)

uneven peaks as seen for standard capillaries in Fig. 3), the fluorescent molecules are sticking to the capillary wall.

9. During capillary scan, the fluorescence signal should be in the range of 100–2000 units. Since all samples contain the same concentration of fluorescently labeled protein, no more than 10% difference in intensity should be seen between capillaries.
10. A lower LED power is preferred to minimize sample photobleaching.

Acknowledgement

This work was supported by the National Institutes of Health grant R01GM121583 awarded to R.R.

References

1. Jerabek-Willemsen M, Wienken CJ, Braun D, Baaske P, Duhr S (2011) Molecular interaction studies using microscale thermophoresis. Assay Drug Dev Technol 9:342–353
2. Seidel SA, Dijkman PM, Lea WA, van den Bogaart G, Jerabek-Willemsen M, Lazic A, Joseph JS, Srinivasan P, Baaske P, Simeonov A, Katritch I, Melo FA, Ladbury JE, Schreiber G, Watts A, Braun D, Duhr S (2013) Microscale thermophoresis quantifies biomolecular interactions under previously challenging conditions. Methods 59:301–315
3. Macdonald PJ, Stepanyants N, Mehrotra N, Mears JA, Qi X, Sesaki H, Ramachandran R (2014) A dimeric equilibrium intermediate nucleates Drp1 reassembly on mitochondrial membranes for fission. Mol Biol Cell 25:1905–1915
4. Ramachandran R, Surka M, Chappie JS, Fowler DM, Foss TR, Song BD, Schmid SL (2007) The dynamin middle domain is critical for tetramerization and higher-order self-assembly. EMBO J 26:559–566
5. Lu B, Kennedy B, Clinton RW, Wang EJ, McHugh D, Stepanyants N, Macdonald PJ, Mears JA, Qi X, Ramachandran R (2018) Steric interference from intrinsically disordered regions controls dynamin-related protein 1 self-assembly during mitochondrial fission. Sci Rep 8:10879
6. Mehrotra N, Nichols J, Ramachandran R (2014) Alternate pleckstrin homology domain orientations regulate dynamin-catalyzed membrane fission. Mol Biol Cell 25:879–890
7. Stowell MH, Marks B, Wigge P, McMahon HT (1999) Nucleotide-dependent conformational changes in dynamin: evidence for a mechanochemical molecular spring. Nat Cell Biol 1:27–32
8. Leonard M, Song BD, Ramachandran R, Schmid SL (2005) Robust colorimetric assays for dynamin's basal and stimulated GTPase activities. Methods Enzymol 404:490–503

Chapter 8

Nucleotide-Dependent Dimerization and Conformational Switching of Atlastin

John P. O'Donnell, Carolyn M. Kelly, and Holger Sondermann

Abstract

A common feature of dynamin-related proteins (DRPs) is their use of guanosine triphosphate (GTP) to control protein dynamics. In the case of the endoplasmic- reticulum- (ER)-resident membrane protein atlastin (ATL), GTP binding and hydrolysis result in membrane fusion of ER tubules and the generation of a branched ER network. In this chapter, we describe two independent methods for dissecting the mechanism underlying nucleotide-dependent quaternary structure and conformational changes of ATL, focusing on size-exclusion chromatography coupled with multi-angle light scattering (SEC–MALS) and Förster resonance energy transfer (FRET), respectively. The high temporal resolution of the FRET-based assays enables the ordering of the molecular events identified in structural and equilibrium-based SEC–MALS studies. In combination, these complementary methods report on the oligomeric states of a system at equilibrium and timing of key steps along the enzyme's catalytic cycle. These methods are broadly applicable to proteins that undergo ligand-induced dimerization and/or conformational changes.

Key words Atlastin, Dynamin-related protein, Oligomerization, Conformational changes, Static light scattering, Förster resonance energy transfer, Enzyme kinetics

1 Introduction

Atlastins (ATLs) belong to the family of dynamin-related proteins (DRPs) that include mechanochemical enzymes that catalyze membrane fission or fusion [1]. The ATL proteins are anchored primarily to the membranes of the reticular endoplasmic reticulum (ER) [2], where they act as homotypic fusogens between tubules, giving rise to the characteristic branching of the ER network [3–5]. Consequently, loss of ATL function results in a reduction of the number of three-way junctions within the peripheral ER. In humans, this cellular phenotype is linked to the expression of mutant, often dominant-negative, ATL alleles that cause hereditary spastic paraplegia (HSP) and hereditary sensory neuropathy

John P. O'Donnell and Carolyn M. Kelly contributed equally to this work.

Rajesh Ramachandran (ed.), *Dynamin Superfamily GTPases: Methods and Protocols*, Methods in Molecular Biology, vol. 2159, https://doi.org/10.1007/978-1-0716-0676-6_8, © Springer Science+Business Media, LLC, part of Springer Nature 2020

(HSN), neurodegenerative diseases affecting motor and sensory neurons, respectively [5–7].

A defining feature of dynamin superfamily members is their ability to hydrolyze GTP and harness that energy to undergo changes in conformation and oligomeric state, which ultimately remodels biological membranes [8]. Common structural features of dynamin-related proteins include a large GTPase domain (G domain), a stalk-like helical domain (middle or M domain), and a transmembrane or membrane-associating domain [1, 9]. Early evidence from studying dynamin defined a general mechanism in which protein monomers are recruited to the neck of clathrin-coated pits at the plasma membrane, where they oligomerize [10, 11]. Coordinated GTP hydrolysis then triggers a series of conformational changes that result in the release of the vesicles from the plasma membrane [8, 9, 12–16]. This mechanism is conserved in the prototypical dynamin and likely applies also to other membrane fission-mediating DRPs [8, 17]. Structural and biochemical studies on guanylate- binding protein 1 (GBP1), another DRP, established that this protein homodimerizes through its G domains upon GTP binding, which establishes its catalytically competent state, preceding conformational changes triggered by GTP hydrolysis [18–20], a mechanism that is conserved in dynamin [21].

Other dynamin-like proteins, including ATL, mediate homotypic membrane fusion (reviewed in refs. 22, 23): mitofusin acts on the outer mitochondrial membrane, while optic atrophy 1 (OPA1) fuses inner mitochondrial membranes. The DRPs involved with membrane fusion function by forming dimers, presumably across opposing membranes, harnessing the energy from GTP hydrolysis to bring membranes in close-enough proximity to mediate fusion [23–27]. The underlying conformational changes have been studied for ATL, using the soluble cytosolic domain as a proxy for the full-length protein [28–32]. The cytosolic module, which we refer to as the catalytic core fragment since it is necessary and sufficient for GTPase activity and dimerization, comprises the DRP's G and helical middle domains. This fragment can be used as an inhibitor for ATL-mediated membrane fusion as it engages with the full-length protein in a GTPase-dependent manner and, thus, prevents productive membrane fusion [33]. On its own, when artificially localized to membranes, it mediates vesicle tethering, which also requires continuous GTP hydrolysis [34], suggesting that the role of the catalytic core is to bring two opposing membranes into close proximity, while fusion also requires the transmembrane domain [35] and a C-terminal amphipathic helix [33, 35, 36]. This helix dips into the cytosolic bilayer leaflet, thinning the membrane and enabling lipid mixing of two bilayers [36].

For ATLs, several crystal structures have revealed distinct states of the catalytic core fragment [29–31]. However, defining how

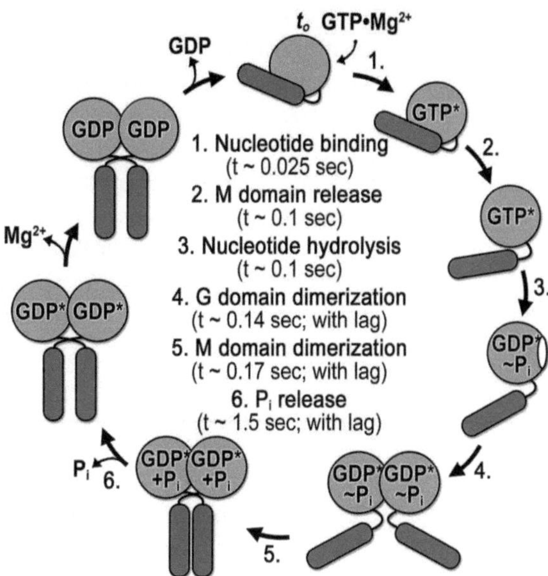

Fig. 1 Kinetic model and sequence of key steps in ATL's GTPase cycle. Pre-steady-state measurements (with high temporal resolution recorded under similar experimental conditions) of individual kinetic steps indicate a clear sequence of events: nucleotide binding, intramolecular conformational changes, and GTP hydrolysis occur in succession, without lag phases; G and M domain dimerizations and phosphate release follow with increasingly long lags. Timing steps shown are based on experiments with human ATL1 core fragment. Modified from [31]

they relate to functional states along ATL's catalytic cycle required kinetics in conjunction with the structural insight. The emerging model comprises the following ordered steps (Fig. 1): (1) GTP loading is facilitated by middle domain docking at the G domain. (2) GTP binding and hydrolysis within monomeric ATL are accompanied by a conformational change, which release the middle from the G domain. (3) After a lag phase, G domain dimerizes, followed closely by middle-domain dimerization, together forming a tight tether that likely drives membrane fusion. (4) Phosphate release proceeds with the longest lag phase, indicating that this step occurs in and likely requires ATL dimers. (5) A crystal structure of hATL3 indicated an intramolecular arginine residue in facilitating the release of the catalytic magnesium prior to GDP release, a step that is important to reset the enzyme for following rounds of GTP hydrolysis [31].

The approaches taken to establish the order of events in ATL's hydrolytic cycle are broadly applicable to other systems. There are two fundamental aims to guide such studies: (1) Determining the effect of different nucleotides on the protein's conformation and/or oligomeric state, and (2) characterizing the timing of the nucleotide-induced events relative to the type of nucleotide and catalysis of the system. In this chapter, we will detail techniques that we have optimized for delineating conformational changes and dimerization, both in steady-state and pre-steady-state regimes. In particular, absolute molecular weight determination by size-exclusion chromatography coupled with multi-angle light scattering (SEC–MALS) is used to describe nucleotide-dependent protein oligomerization. Förster resonance energy transfer- (FRET-) based assays using rapid-mixing, stopped-flow analysis is used to monitor both intramolecular conformational changes and domain oligomerization.

2 Material

2.1 Size-Exclusion Chromatography Coupled with Multi-angle Light Scattering (SEC–MALS)

1. Size-exclusion chromatography column (or HPLC column); in this step, we use Superdex 200 Increase 10/300 GL (GE Healthcare).
2. HPLC system (Agilent Technologies) with UV detector.
3. Light scattering and refractive index detectors (DAWN HELEOS II and Optilab® T-rEX™, Wyatt Technology).
4. MALS buffer: 25 mM Tris at pH 7.5, 100 mM NaCl, 4 mM $MgCl_2$, and 2 mM EGTA (*see* **Note 1**).
5. Monomeric fraction of bovine serum albumin (BSA) (Sigma: A1900).
6. Nucleotides (Sigma: G7127, GDP; G8634, GTPγS; and G0635, GppNHp) reconstituted in 100 mM Tris, pH adjusted to 7.5, and stored at −20 °C (10–20 μL aliquots).

2.2 Designing FRET Expression Constructs for Sulfhydryl-Reactive Chemistries

1. Alignment tool for selecting non-conserved residues of interest and structure coordinates for protein of interest if available (e.g., Clustal Omega: https://www.ebi.ac.uk/Tools/msa/clustalo/ and https://www.rcsb.org).
2. PyMOL molecular visualization software (Schrödinger: https://pymol.org/2/).
3. QuikChange II Site-Directed Mutagenesis Kit (Agilent: 200523).

2.3 Functional Validation of FRET-Engineered Protein

1. EnzChek™ Phosphate Assay Kit (Thermo Fisher: E6646), Malachite Green Phosphate Assay Kit (Sigma: MAK307), or Phosphate Sensor (Thermo Fisher: PV4407).
2. Size-exclusion chromatography coupled with multi-angle light scattering (SEC–MALS) (*see* Subheadings 2.1 and 3.1).

2.4 Maleimide–Sulfhydryl Labeling Reaction

1. Labeling buffer: 25 mM HEPES at pH 7.0 and 100 mM NaCl (*see* **Note 2**).
2. Alexa Fluor™ 488 C5 Maleimide (AF-488) and Alexa Fluor™ 647 C2 Maleimide (AF-647) (Thermo Fischer: A10254 and A20347). Reconstitute dyes in dimethyl sulfoxide (DMSO) to 10 mM, aliquot into 10 μL volumes, and store at −20 °C.
3. illustra™ NAP-5 Column (GE: 17085301).
4. Reaction buffer: 25 mM HEPES pH 7.5, 100 mM NaCl, and 2 mM $MgCl_2$.
5. NanoDrop™ 2000/2000c Spectrophotometers (ND-2000/ND2000c).
6. SDS-PAGE gel running/staining equipment.
7. In-gel fluorescence imager (e.g., Bio-Rad Gel Doc system) with LED excitation and emission filters for desired fluorophores used in labeling.
8. Gel quantification software, such as, Bio-Rad Image Lab™ Software or Fiji (https://fiji.sc).
9. Data analysis and graphing software (e.g., GraphPad Prism: https://www.graphpad.com or SigmaPlot: http://sigmaplot.co.uk/products/sigmaplot/sigmaplot-details.php).

2.5 Stopped-Flow Experiment and Data Processing

1. Stopped-flow spectrometer (Applied Photophysics SX20™ or KinTek Corporation SF-300x).
2. Emission filters: Chroma D525/50 (donor channel) and SCHOTT 645 nm high-pass filter (accepter channel).
3. GTP (100 mM aqueous solution; Thermo Fischer: R0461).
4. Data analysis and graphing software (e.g., GraphPad Prism: https://www.graphpad.com or SigmaPlot: http://sigmaplot.co.uk/products/sigmaplot/sigmaplot-details.php).

3 Methods

3.1 Size-Exclusion Chromatography Coupled with Multi-angle Light Scattering (SEC–MALS)

SEC–MALS experiments are used to determine the molecular weight of ATL proteins experimentally trapped at various stages of the GTP-hydrolysis cycle (Fig. 2). When considering five distinct states (apo, GTP-bound, transition, post-hydrolysis, and GDP-bound), there are several approaches to decipher how nucleotide binding and/or hydrolysis influences oligomerization. The use of non-native nucleotides can be a powerful tool to mimic specific states in the catalytic cycle. These non-native nucleotides generally fall into two categories with GTPγS, GppNHp, GppCHp, and GDP·BeF_3^- being considered non-hydrolyzable analogues or mimics of GTP and GDP·AlF_4^- being considered a transition-state analogue [37, 38]. Despite these commonly accepted categories,

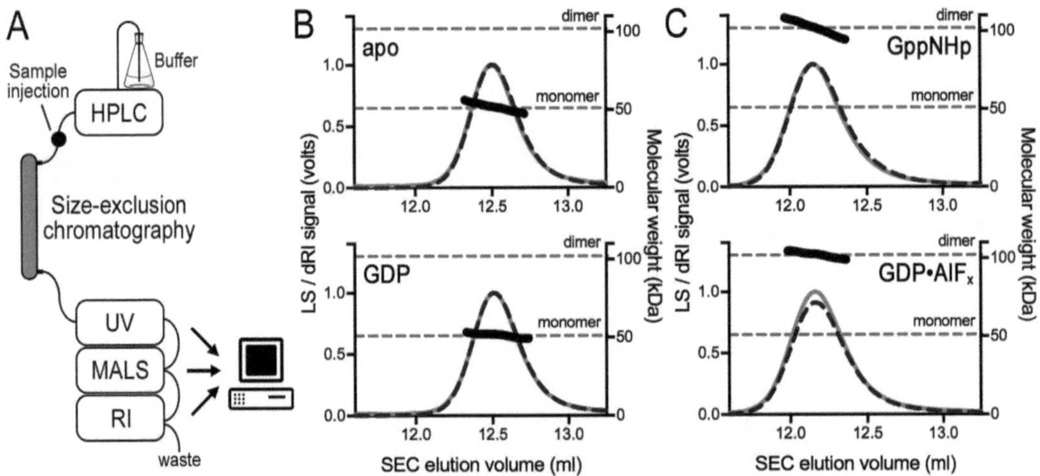

Fig. 2 SEC–MALS data collection and analysis. (**a**) Diagram of the SEC–MALS setup. (**b**) Sample data for monomeric human ATL1 catalytic core fragment in the absence of nucleotide or preincubated with GDP·Mg^{2+}. (**c**) Sample data for dimeric human ATL1 catalytic core fragment preincubated with GppNHp·Mg^{2+} or GDP·AlF_4^-·Mg^{2+}·EGTA [31]

how the electronic structures of non-native nucleotides affect the range of ligand-induced conformational states can be G protein specific. Caution should be taken when inferring hydrolysis mechanisms from the use of nucleotide mimetic. A complimentary approach relies on the native substrate GTP in combination with catalytically inactive, yet substrate binding-competent, mutant variants of the protein (e.g., human ATL1-R^{77}A). Both these approaches are useful for experiments that require prolonged or trapped conformational states. However, the studies on ATL showed that prolonged incubations with GTP analogues or GTP in the absence of catalysis allow the system to equilibrate to a common conformation, convoluting the structural analysis of ATL's functional cycle [29–31]. Thus, SEC–MALS experiments are useful to establish oligomeric states of ATL relative to the nucleotide bound, which set the stage for pre-steady-state kinetics revealing events as a consequence of GTP hydrolysis at physiologically relevant time scales (*see* Subheadings 3.2–3.7).

1. System Equilibration:
   (a) Attach size-exclusion chromatography column to HPLC and equilibrate in MALS buffer according to the manufacturer's recommendations with a flow rate and backpressure compatible with the column.
   (b) Prepare light-scattering detector by turning on the flow-cell sonicator (turn "COMET" to "ON") (*see* **Note 3a**).

(c) Prepare refractometer by opening reference flow cell to replace storage buffer with MALS buffer (turn "PURGE" to "ON") (*see* **Note 3b**).

(d) Run until light-scattering and refractive index signals are baselined (about 1.5 h). Turn off COMET and close the PURGE valve at least 15 min before running samples.

2. Prepare Samples:

(a) Prepare 100 μL samples of 40 μM ATL catalytic core fragment (e.g., residues 1-446 of human ATL1) and add nucleotide (GDP, GppNHp, GTPγS, or GDP·AlF_4^-) to a final concentration of 2 mM (*see* **Note 4**).

- Incubate for 1 h before sample analysis to allow equilibrium to be reached.
- Spin all samples at 16,000 × *g* for 10 min before loading to remove precipitate.

(b) Include a sample of BSA monomeric fraction as a standard; reconstitute at 5 mg/mL in MALS buffer.

3. Running samples:

(a) Load 40 μL sample and run each condition for 1.25 column volumes to reestablish baselines between runs.

(b) Use Astra software (Wyatt Technology) to collect data for light–scattering and refractive index and UV at 280 nm; data points are collected for each measurement every 0.5 s.

4. Data analysis.

(a) Molecular weight of BSA is calculated using Wyatt's Astra software (*see* **Note 5**). The formula

$$I(\theta)_s \propto Mc\left(\frac{dn}{dc}\right)^2 \tag{1}$$

describes the intensity of light scattering as a function of angle, $I(\theta)_s$, which is proportionate to molecular mass of the scatterer (M), protein concentration (c), and square of the specific refractive index increment (dn/dc) (Eq. 1).

(b) Both configuration and method files from the BSA standard are applied to all sample data files. M_W (weight-averaged molecular mass, used as absolute molecular mass) is calculated every 0.5 s within specified ranges corresponding to peak elution regions. Measurements are usually averaged across the peak (*see* **Note 6**).

(c) For each selected peak, determine monodispersity using the polydispersity index, where a value of "1" indicates a monodisperse sample.

3.2 Designing FRET Expression Constructs for Sulfhydryl-Reactive Chemistries

Because DRPs often homodimerize and undergo conformational changes, FRET experiments can be designed to monitor intermolecular interactions or intramolecular switching using a singly or doubly labeled protein, respectively [39]. Experimental design includes the selection of one (intermolecular FRET) or two (intramolecular FRET) surface-exposed residue(s) for labeling, preferentially using structural information as a guide. The chosen residues should not be conserved and should be surface exposed to promote efficient labeling, and the expected distance changes between labeled residues should sample the Forster distance (R_o) for a given FRET donor/acceptor pair (~56 Å for AF-488/AF-647) (*see* **Note 7**). In the protocol described in this chapter, site-specific labeling of the target protein(s) is achieved through covalent modification of cysteine thiols via maleimide-containing fluorophores; however, alternative approaches that generate site-specific labeled proteins can be employed. Generally, surface-exposed cysteine residues other than the desired labeling sites must be mutated to a nonreactive residue to allow for one (intermolecular FRET) or two (intramolecular FRET) unique fluorophores per protein molecule. In this chapter, we will outline a strategy to measure intermolecular FRET where two monomeric proteins homodimerize to produce a FRET signal. Despite leaving the following protocols general for application to any target protein, these strategies were used to develop G domain and middle domain FRET assays for both hATL1 and hATL3 [29, 31].

1. Using Clustal Omega, align five to ten protein sequences, include the target protein and target protein sequences from closely related organisms. If the degree of conservation is very high, sequences of more distantly related orthologs can be included.
2. Notate non-conserved residues in context of their surface availability. Surface availability can be easily determined if a three-dimensional structure of the target protein is available in the Protein Data Bank (PDB). Using the molecular visualization software PyMOL, manually determine which of the non-conserved residues are on the surface of the protein. If no structural information is available, alternative approaches can be used (*see* **Note 8**).
3. If structural models exist for two or more conformations to be investigated via FRET, use PyMOL to calculate the distance between the potential non-conserved residues available on the surface in each state. Select three pairs of non-conserved residues with the largest distance change between conformations.
4. Determine the number of protein cysteine residues the target protein contains and their surface availability similar to the above strategies for non-conserved residues. Mutate the

cysteine residues to either alanine or serine depending if the environment of the native cysteine side chain is hydrophobic or hydrophilic. If a large number of cysteine residues are present, mutating a subset of the cysteines or an alternative labeling strategy may be used (*see* **Note 9**). Use QuikChange II (or QuikChange Multi-) Site-Directed Mutagenesis with the protocols developed by the manufacturer for generating cysteine-less (cys-less) construct.

5. Select the top three FRET residue pairs and mutate the residues to cysteine using QuikChange II Site-Directed Mutagenesis.
6. The FRET expression constructs can be expressed and the proteins can be purified according to published methods [29, 31].

3.3 Functional Validation of FRET-Engineered Protein

Determining that FRET-destined proteins' function the same as wild-type protein is an essential control. It is recommended to determine the effect of mutations on protein activity throughout the process of generating FRET constructs to keep track of how various mutants behave.

1. Conduct a phosphate-release assay to determine the turnover number (or equivalent assay to compare enzymatic activity). EnzChek™ Phosphate Assay Kit (Thermo Fisher: E6646), Malachite Green Phosphate Assay Kit (Sigma: MAK307), or Phosphate Sensor (Thermo Fisher: PV4407) can be used in accordance with the manufacturer's recommendations (*see* **Note 10**).
2. In addition to activity, nucleotide-dependent oligomerization can be used to judge functional protein. In this chapter, SEC–MALS can be used as described above (*see* Subheading 3.1)

3.4 Small-Scale Maleimide–Sulfhydryl Labeling Optimization

Ideally, labeling reaction optimization should yield maximum specific labeling while mitigating any nonspecific labeling. The duration of protein labeling is an important parameter to change as it can vary between proteins and the labeling sites selected. Additional parameters that can be varied are protein:dye ratio, protein concentration, and temperature. The optimization protocol below uses small volumes and omits a buffer exchange step. This is enabled by the use of in-gel fluorescence but requires access to a Bio-Rad ChemiDoc™ or equivalent imaging system to quantify the level of dye incorporation (Fig. 3). Alternatively, calculating the labeling efficiency of a buffer-exchanged, labeled protein as described in Subheading 3.5 will provide similar data.

1. Prepare 100 μL of 2× FRET protein stock solution by diluting the protein- engineered for FRET to 200 μM in labeling buffer; keep on ice.

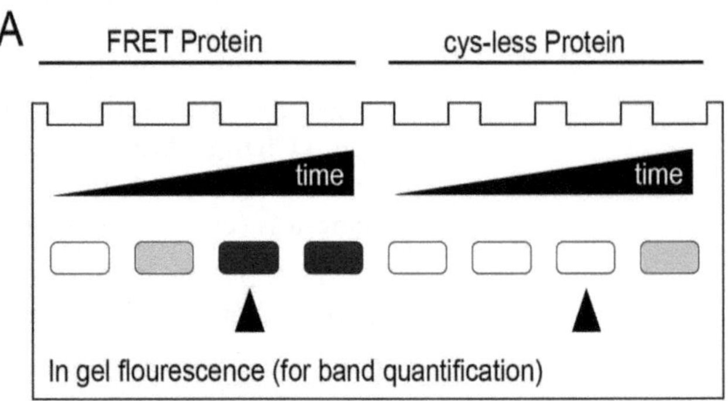

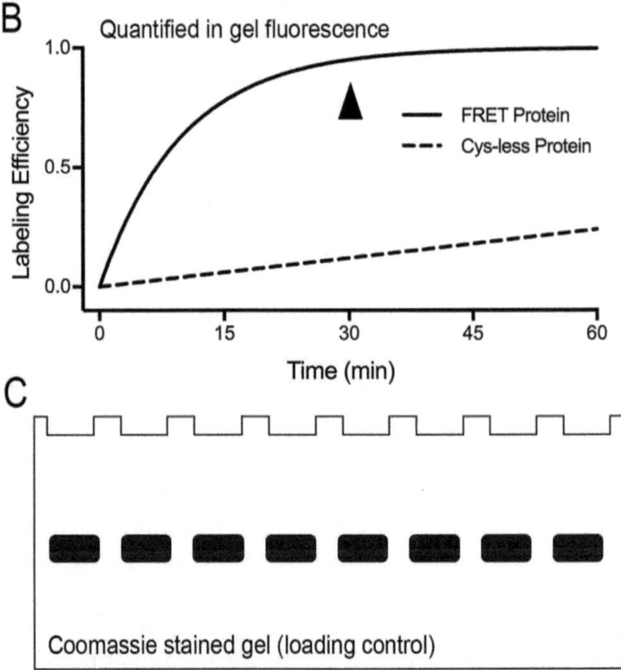

Fig. 3 Optimization of labeling efficiency. (**a**) Individual time points from the labeling reactions can be separated from excess dye using standard gel electrophoresis. The SDS-PAGE gel can be imaged using the intrinsic fluorescence of the labeled protein using a gel imaging system with fluorescence capabilities. Black arrows indicate optimal labeling time where the FRET protein is sufficiently labeled, but the cys-less protein remains unmodified. (**b**) Band fluorescence can be quantified using Bio-Rad Image Lab™ or Fiji and plotted in GraphPad Prism or SigmaPlot. (**c**) The gel can be subsequently stained with Coomassie dye to ensure proper loading equivalents

2. Prepare a 100 μL of 2× cys-less protein stock by diluting cys-less protein to 200 μM in labeling buffer; keep on ice.
3. Prepare 100 μL of 2× donor and 2× acceptor dye stock solutions by adding 3 μL of 10 mM maleimide–fluorophore stock

solutions (AF-488 or AF-647 in DMSO) to 97 μL of labeling buffer in two separate tubes. The final 2× dye concentration is 300 μM for each solution; keep on ice.

4. Combine 45 μL of 2× protein stock and 45 μL of 2× dye stock to initiate the reaction; keep on ice. Initiate reactions for the FRET-engineered proteins and corresponding cys-less protein for both donor and acceptor dyes for the total of four reactions. Remove 10 μL of the reaction and quench with 1 μL of 20 mM DTT (i.e., excess-free sulfhydryls) at various time points (0, 5, 15, 30, 45, 60, and 90 min).
5. Dilute the 10 μL of quenched reaction with 40 μL of 1.25× SDS sample buffer. Run 5 μL of each diluted sample (~5 μg of protein) on an SDS-PAGE gel using standard methods (including a serial dilution of sample on a separate gel is advisable to test the linear range of in gel fluorescence).
6. Image gels by fluorescence using a Bio-Rad ChemiDoc™ XRS + Imager or equivalent gel imaging system that has fluorescence capabilities. Bio-Rad recommends a 470/30 nm excitation filter and 530/28 emission filter for AF-488 and 625/30 nm excitation filter and 695/55 emission filter for AF-647 (Fig. 3a). Band intensities can be calculated using Bio-Rad Image Lab™ Software or Fiji (https://fiji.sc) and plotted as a function of time in GraphPad Prism (Fig. 3b). Select the labeling time that maximizes specific labeling of the introduced cysteine residue and minimizes nonspecific labeling of the cys-less protein (Fig. 3; black arrows).
7. The gel can subsequently be Coomassie stained for a loading control using standard methods (Fig. 3c).

3.5 Large-Scale Maleimide–Sulfhydryl Labeling for FRET Experiments

Once the labeling reaction is optimized, reaction volumes can be scaled up to produce labeled protein sufficient for subsequent stopped-flow experiments.

1. Prepare 260 μL of 2× FRET protein stock solution by diluting protein engineered for FRET to 200 μM in labeling buffer; keep on ice.
2. Prepare 125 μL 2× donor dye stock by adding 3.75 μL of 10 mM AF-488 and 121.25 μL of labeling buffer, and prepare 125 μL 2× acceptor dye stock by adding 3.75 μL of 10 mM AF-647 and 121.25 μL of labeling buffer for a final dye concentration of 300 μM; keep on ice.
3. Add 125 μL of the 2× FRET-destined protein stock solution to the tubes containing 125 μL 2× donor dye; let reaction progress for the optimal time determined above.

4. Add 125 μL of the 2× FRET-destined protein stock solution to the tubes containing 125 μL 2× acceptor dye; let reaction progress for the optimal time determined above.
5. Rather than quenching with DTT at the end of the reaction, proceed directly to a buffer exchange step using illustra™ NAP-5 columns using the standard manufacturer's protocols. Buffer exchange will quench the reaction by removing excess dye.
6. Determine the efficiency of labeling using a NanoDrop™ 2000/2000c Spectrophotometer.
   (a) Measure the protein concentration using absorbance at 280 nm, a theoretical extinction coefficient calculated using ProtParam (https://web.expasy.org/protparam/), and the sequence of the purified protein as the input.
   (b) Measure the dye concentration. AF-488 has a maximum absorbance at 496 nm with an extinction coefficient of 71,000 cm^{-1} M^{-1}. AF-647 has a maximum absorbance at 650 nm with an extinction coefficient of 239,000 cm^{-1} M^{-1}.
   (c) Divide the dye concentration by the protein concentration to determine the fraction of labeled protein. If labeling efficiency is less than 0.5 or greater than 1, alternative residues should be investigated for labeling. For the G domain of ATL, yields of 0.95 labeling ratio were achieved routinely.

3.6 Stopped-Flow Experiment Setup and Controls

The stopped-flow experiment measuring intermolecular FRET between G domains of human ATL3 was optimized using an Applied Photophysics SX20™, although any stopped-flow with fluorescence capabilities can be used (Fig. 4a).

1. Prepare the SX20 stopped-flow device in accordance of the user manual. These steps include:
   (a) Turn on nitrogen supply to a pressure of 8 bar.
   (b) Power on electronics unit.
   (c) Power and ignite the 150 W xenon arc lamp.
   (d) Power on computer and load Pro-data SX20 software.
   (e) Adjust stop syringe to desired volume (60 μL of each reaction solution resulting in 120 μL total stop volume).
   (f) Adjust the excitation wavelength using the monochromator to 473 nm (*see* **Note 11**).
   (g) Install the donor filter into PMT1.
   (h) Install the emission filter into PMT2 (*see* **Note 12**).
   (i) Flush the stopped-flow unit with 3 mL water.
   (j) Flush the stopped-flow unit with 3 mL reaction buffer.

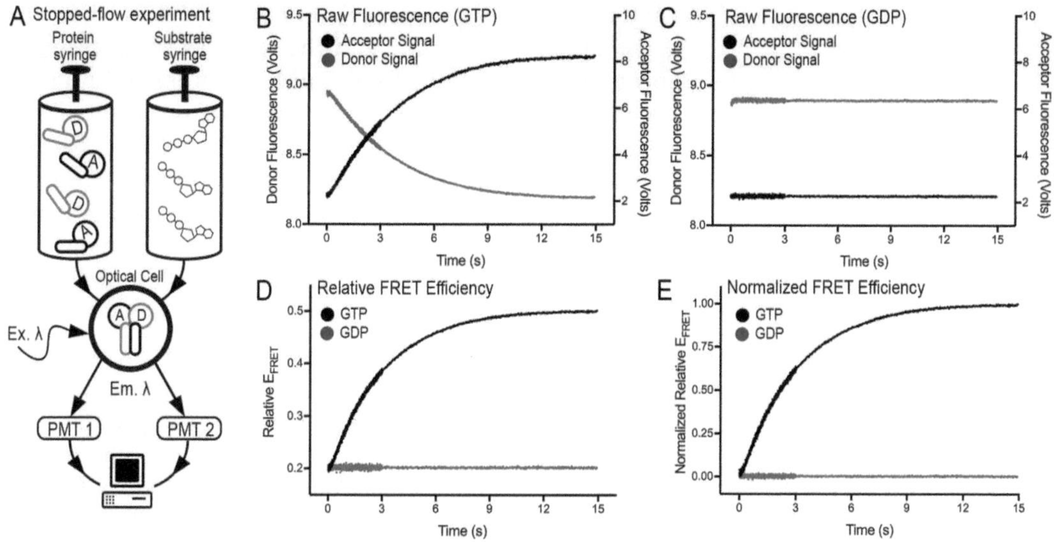

Fig. 4 FRET assay. (**a**) Schematic of a stopped-flow FRET experiment. The protein drive syringe is loaded with both donor (D, gray) and acceptor (A, black) forms of ATL, and the substrate drive syringe is loaded with GTP. Upon injection, the contents of the two drive syringes are mixed, and the reaction is left to age in the optical cell. In this experiment, the fluorophores are excited with a xenon lamp where a monochromator selects the desired excitation wavelength. Photons emitted from both donor and acceptor fluorophores are detected with two photomultiplier tubes (PMTs), and these voltages are recorded by the instrument's software. (**b**) Raw FRET data using GTP as substrate. Donor (gray, left axis) and acceptor (black, right axis) FRET signals for an experiment monitoring GTP hydrolysis-dependent dimerization of human ATL3's GTPase domains within the catalytic core fragment. Upon dimerization, there is a loss in donor fluorescence and a concomitant increase in acceptor fluorescence. (**c**) The same experiment as in **a**, but with GDP, a nucleotide that does not support homodimerization. (**d**) A relative FRET efficiency can be calculated (*see* Eq. 2) from the donor and acceptor signals, and FRET experiments from different nucleotides can be compared. GTP is shown in black and GDP is shown in gray. (**e**) Data sets can be normalized to the experiment yielding the large change in relative FRET efficiency (*see* Eq. 3). This preserves reaction rates and the ability to compare amplitudes of different nucleotides (GTP, black; GDP, gray)

2. The following 2× samples should be made in reaction buffer for a final volume of 2 mL for initial controls:
   (a) 2× Protein: FRET pair (1 μM donor + 1 μM acceptor)
   (b) 2× Protein: donor only (1 μM donor only)
   (c) 2× Protein: acceptor only (1 μM acceptor only)
   (d) 2× Substrate: GTP (1 mM GTP)
   (e) Reaction buffer only ("apo").
3. The PMT high voltage should be set so that the expected photon flux for a given experiment does not overload the PMT. This should be done for the donor and acceptor PMTs using the low- and high-FRET states, respectively.
   (a) First, set the donor emission PMT by running the reaction where the lowest- FRET state is expected (e.g., the

highest donor fluorescence). For ATL, this is when the proteins are monomeric and the reaction includes 1 μM protein (0.5 μM donor + 0.5 μM acceptor) and no substrate in the optical cell. Add 1 mL of the protein and buffer solutions separately to each of the two drive syringes and click "drive" 8–10 times to ensure the protein/buffer solutions fully occupy the optical cell. Click "auto PMT" in the software for the PMT that contains the donor emission filter. Flush the stopped-flow unit with 3 mL reaction buffer.

(b) Next, one can define the high voltage for the acceptor emission PMT using the reaction for which the highest-FRET state is expected (e.g., the highest acceptor fluorescence). For ATL, this is when the proteins are dimeric and the reaction includes 1 μM protein (0.5 μM donor + 0.5 μM acceptor) and 500 μM GTP in the optical cell. As above, flush the stopped-flow unit with 8–10 drives of the protein and GTP solutions. Wait ~0.5–1 min for the reaction to reach the high-FRET state before substrate is depleted (this time can be calculated from the enzyme's turnover number), and click "auto PMT" in the software for PMT2, which contains the appropriate acceptor emission filter. Flush the stopped-flow unit with 3 mL reaction buffer.

(c) Record these voltages for PMT1 and PMT2.

4. Controls should be conducted to determine if there is any cross-contamination between the donor and acceptor channels in the experimental setup. Two main scenarios can result in channel contamination. The acceptor fluorophore can be directly excited at the donor excitation wavelength or donor emission signal can bleed-through the acceptor emission filter. Controls for fluorescence channel cross talk can be tested by measuring fluorescence of the reaction buffer at the PMT voltages determined above in both the donor and acceptor channels while exciting at 473 nm. The baseline buffer fluorescence can be compared to two controls: 1 μM donor only or 1 μM acceptor only in the optical cell (*see* **Note 13**).

3.7 Stopped-Flow Experiment and Data Processing

With the instrumentation setup and controls completed, experimental data can be collected. There are many data collection settings and parameters that can be controlled in the Pro-data SX20 software. It is suggested to start with the manufacturer's default settings and consult the SX20 user manual to determine how to make specific modifications. Settings that have been found beneficial are the pressure hold option that eliminates an instrument artifact at ~30 ms (only used for collections <2 s) and split time base, which allows the data collection points to be distributed across two linear time intervals.

1. For a given experiment, prepare 2 mL of 2× protein stock solution that contains 1 μM donor-labeled protein and 1 μM acceptor-labeled protein, as well as 2 mL of 2× substrate stock solution that contains 1 mM GTP (or other nucleotide to be tested).
2. Fill each drive syringes with either protein or substrate (Fig. 4a); click drive 6–8 times to flush the system.
3. Set the data collection parameters. In this chapter, we collected data for 15 s, using a split time base with 1000 data points for the first 3 s and 500 data points for the remaining 12 s. Data can then be acquired and exported.
4. Upon collecting experimental data, stopped-flow device can be flushed with 6 mL of water, powered off, and nitrogen turned off for the next user.
5. The number of technical replicates required will depend on the individual experiment and the comparisons being made. It is suggested to have at a minimum six technical replicates and three biological replicates from separate protein purifications and labeling reactions.
6. Comparing the donor and acceptor fluorescence channels, homodimerization should result in a loss of donor signal and a concomitant increase in acceptor signal. This is the case when ATL is mixed with GTP, which initiates homodimerization (Fig. 4b), but not when mixed with GDP, which does not support dimerization (Fig. 4c).
7. In the case of ATL, particular questions pertain to differences in the rates of homodimerization between domains and nucleotides. For this reason we calculate a relative FRET efficiency.

$$E_{\mathrm{FRET}} = \frac{I_A}{I_A + I_D} \tag{2}$$

   where I_A and I_D are the intensities of the acceptor and donor channel, respectively (*see* **Note 14**). When both donor and acceptor signals are measured from the same reaction (e.g., two PMTs are used for data collection), relative FRET efficiencies are highly reproducible. It is recommended to always calculate relative FRET efficiencies for individual stopped-flow replicates as the donor and acceptor channels are paired (Fig. 4d). Relative FRET efficiencies from individual technical and biological replicates can be averaged.

8. Though not necessary, the data may benefit from a normalized representation to make certain comparisons. When

normalizing relative FRET efficiencies, it is important to remember that though rate constants will not be impacted by normalizing individual data sets, the heights of the curves (e.g., relative FRET efficiency) may change, making it impractical for comparing results between different nucleotides in a quantitative manner. To preserve all information and quantitatively compare FRET efficiencies for reactions triggered with different nucleotides, it is suggested to normalize all data sets to the GTP data set.

$$z_i = \frac{y_i - \min(y)}{\max(y) - \min(y)} \tag{3}$$

where $y = (y_1, \ldots, y_n)$, $\min(y)$ and $\max(y)$ are the minimum and maximum y values in the GTP data set, and z_i is the ith normalized data point (Fig. 4e).

4 Notes

1. Consider what conditions are required for the protein under investigation to bind nucleotide. ATL requires $MgCl_2$ for any nucleotide binding and additional EGTA for binding AlF_4^-. If nucleotide-induced complexes show signs of disassembly upon dilution during gel filtration, consider including nucleotides or nucleotide analogues in the mobile phase.
2. The maleimide moiety used here for bioconjugation to cysteine sulfhydryl groups is most specific at a pH between 6.5 and 7.5. Under more basic conditions (pH > 8.5), the reaction begins to favor primary amines (e.g., lysine residues) and increased rates of maleimide hydrolysis (Crosslinking Reagents Technical Handbook: https://tools.thermofisher.com/content/sfs/brochures/1602163-Crosslinking-Reagents-Handbook.pdf). HEPES buffers are suited for labeling reactions as their pH is independent of temperature, unlike other common buffers, such as, Tris base.
3. (a): The COMET flow cell sonicator is used to clean any potential contaminants from previous samples or buffers. Even small amounts can impact light scattering signal. (b): The PURGE opens the reference flow cell to replace the system storage buffer with MALS buffer. Differential refractive index measurements require an exact buffer match to sample flow cell.
4. To set up the AlF_4^- condition, first prepare 2 mM GDP, then add 2 mM $AlCl_2$, and 20 mM NaF and mix by pipetting.
5. BSA is used as a standard because it is an isotropic scatterer. BSA has a known mass and diameter that is sufficiently small

relative to the wavelength of the laser light used in the detectors so that scattering intensities do not show angular dependence. These properties render BSA ideal for the calibration and normalization of all light scattering detectors relative to the signal at the 90° detector. This is flow-cell specific, so calibration from one instrument cannot be used to analyze data from another instrument.

6. Size-exclusion chromatography is only used as a fractionation step and the peak elution volume is not a factor in the molecular mass calculation. Consequently, a difference in elution volume of the same protein under different conditions (e.g., nucleotide-bound state) without a change in the measured molecular masses across the elution peak is indicative of conformational changes due to distinct hydrodynamic radii of the two species. Molecular masses that are calculated to be intermediate between oligomeric species (e.g., between a monomer and a dimer) typically arise when the proteins dissociate during the course for the chromatography step and/or undergo fast association and dissociation relative to the measurement timescale. Often, this phenomenon is associated with a trend of molecular masses measured across the peak, with higher values at the beginning of the peak. In these cases, molecular mass determinations often show sample concentration dependence, with higher concentrations biasing the population toward the higher oligomeric species and lower concentrations favoring the lower oligomeric species. If the rates of exchange for a given interaction are slow relative to the measurement timescale, a bimodal (or multimodal) peak distribution can be expected.

7. For ATL, intermolecular FRET experiments yielded robust signal as the starting distance was infinite (e.g., monomeric species), and homodimerization resulted in final distances between 30 Å and 40 Å. Note that FRET signal can be optimized by using FRET dye pairs that have an R_o value that is closest to the expected mean distance change. For example, if the starting distance is 30 Å and the ending distance is 70 Å, a dye pair with R_o of 50 Å would be ideal; donor/acceptor pairs that could be used include AF-350/AF-488 and AF-555/AF-647. Refer to The Molecular Probes Handbook (http://thermofisher.com/handbook) and Fluorescence SpectraViewer (http://thermofisher.com/spectraviewer) as the starting point to find information pertaining to common FRET pair R_o values and excitation/emission wavelengths of Alexa Fluor™ dyes.

8. If a structure for the protein of interest is not available, use BLASTp while restricting the search query to the Protein Data Bank (PDB) database to determine if structural information

exists for closely related protein homologues based on sequence (https://blast.ncbi.nlm.nih.gov/Blast.cgi?PAGE=Proteins). If no structural information can be attained, surface accessibility can be predicted using computational methods, which include SARpred or RSARF [40, 41].

9. Often, not all cysteine residues require mutation to ensure a singly labeled protein. In the case of human ATL3, the catalytic core fragment (residues 1–442) contains seven cysteines, but mutation of the four surface-exposed cysteines yields a construct that is resistant to nonspecific labeling. If there are too many cysteines or mutation of surface-exposed cysteines impedes protein function, AMBER suppression technology can be applied to introduce bio-orthogonal amino acids that are amenable to labeling [34, 42, 43]. Though restricted to the N- and C-termini, ECFY and EYFP can be genetically fused to the target protein to measure FRET changes in specific cases [29].

10. There are multiple ways of assaying phosphate release, and each assay has benefits and drawbacks. EnzChek™ Phosphate Assay Kit (Thermo Fisher: E6646) is a multi-turnover coupled reaction where inorganic phosphate is measured by absorption in real time. Since the inorganic phosphate is consumed by the coupled reaction, reaction rates can be artificially higher due to lack of product inhibition. Malachite Green Phosphate Assay Kit is an end point absorption assay with no coupled enzymes but requires individual time points to be taken (Sigma: MAK307). Phosphate Sensor (Thermo Fisher: PV4407) is a real-time single-turnover fluorescence reaction which is the most sensitive assay (other than using radioactive- ^{32}P-labeled GTP).

11. It is important to optimize the excitation wavelength for a given experiment. Though the Molecular Probes Handbook recommends using an excitation wavelength of 473 nm, we have successfully used wavelengths between 420 nm and 473 nm depending on the experimental setup and hardware being used. These values can be determined empirically by conducting excitation and emission scans in a standard cuvette-based fluorometer.

12. We find that the standard Hamamatsu R6095 PMT is right on the cusp of detecting the emission of AF-647. To increase sensitivity, we recommend using a 645 nm SCHOTT glass cut-off filter and a red-shifted PMT (Hamamatsu R374), which has a spectral response range between 185 nm and 850 nm.

13. We have found that for many experiments, an excitation value of 420 nm eliminates all acceptor excitation at the donor-

excitation wavelength. However, this short of an excitation wavelength for AF-488 can only be used in the most robust FRET experiments as this wavelength likely excites only ~5% of the AF-488 population. A secondary way of eliminating cross talk is by analytical correction methods. Elder et al. provides a concise, quantitative protocol removing various types of channel contamination [44].

14. We refer to Eq. 2 as a relative FRET efficiency in consideration of the combinations that donor and acceptor proteins can sample during homodimerization (i.e., donor–donor, donor–acceptor, or acceptor–acceptor). Changing the ratio of donor and acceptor-labeled proteins (e.g., 1:20 donor to acceptor) can bias the reaction, so donor-labeled proteins are predominantly paired with an acceptor-labeled protein [29]. This strategy can also be used in intramolecular FRET experiments where an excess of unlabeled protein can be included to mitigate any signal contribution from intermolecular effects [31].

Acknowledgement

This work was supported by the Spastic Paraplegia Foundation to H.S.

References

1. Praefcke GJ, McMahon HT (2004) The dynamin superfamily: universal membrane tubulation and fission molecules? Nat Rev Mol Cell Biol 5(2):133–147
2. Zhu PP, Patterson A, Lavoie B, Stadler J, Shoeb M, Patel R, Blackstone C (2003) Cellular localization, oligomerization, and membrane association of the hereditary spastic paraplegia 3A (SPG3A) protein atlastin. J Biol Chem 278(49):49063–49071
3. Hu J, Shibata Y, Zhu PP, Voss C, Rismanchi N, Prinz WA, Rapoport TA, Blackstone C (2009) A class of dynamin-like GTPases involved in the generation of the tubular ER network. Cell 138(3):549–561
4. Orso G, Pendin D, Liu S, Tosetto J, Moss TJ, Faust JE, Micaroni M, Egorova A, Martinuzzi A, McNew JA, Daga A (2009) Homotypic fusion of ER membranes requires the dynamin-like GTPase atlastin. Nature 460 (7258):978–983
5. Rismanchi N, Soderblom C, Stadler J, Zhu PP, Blackstone C (2008) Atlastin GTPases are required for Golgi apparatus and ER morphogenesis. Hum Mol Genet 17(11):1591–1604
6. Fink JK (2006) Hereditary spastic paraplegia. Curr Neurol Neurosci Rep 6(1):65–76
7. Guelly C, Zhu PP, Leonardis L, Papic L, Zidar J, Schabhuttl M, Strohmaier H, Weis J, Strom TM, Baets J, Willems J, De Jonghe P, Reilly MM, Frohlich E, Hatz M, Trajanoski S, Pieber TR, Janecke AR, Blackstone C, Auer-Grumbach M (2011) Targeted high-throughput sequencing identifies mutations in atlastin-1 as a cause of hereditary sensory neuropathy type I. Am J Hum Genet 88 (1):99–105
8. Daumke O, Praefcke GJ (2016) Invited review: mechanisms of GTP hydrolysis and conformational transitions in the dynamin superfamily. Biopolymers 105(8):580–593
9. Antonny B, Burd C, De Camilli P, Chen E, Daumke O, Faelber K, Ford M, Frolov VA, Frost A, Hinshaw JE, Kirchhausen T, Kozlov MM, Lenz M, Low HH, McMahon H, Merrifield C, Pollard TD, Robinson PJ, Roux A, Schmid S (2016) Membrane fission by dynamin: what we know and what we need to know. EMBO J 35(21):2270–2284
10. Hinshaw JE, Schmid SL (1995) Dynamin self-assembles into rings suggesting a mechanism

for coated vesicle budding. Nature 374 (6518):190–192

11. Takei K, McPherson PS, Schmid SL, De Camilli P (1995) Tubular membrane invaginations coated by dynamin rings are induced by GTP-gamma S in nerve terminals. Nature 374 (6518):186–190
12. Chappie JS, Mears JA, Fang S, Leonard M, Schmid SL, Milligan RA, Hinshaw JE, Dyda F (2011) A pseudoatomic model of the dynamin polymer identifies a hydrolysis-dependent powerstroke. Cell 147(1):209–222
13. Ford MG, Jenni S, Nunnari J (2011) The crystal structure of dynamin. Nature 477 (7366):561–566
14. Marks B, Stowell MH, Vallis Y, Mills IG, Gibson A, Hopkins CR, McMahon HT (2001) GTPase activity of dynamin and resulting conformation change are essential for endocytosis. Nature 410(6825):231–235
15. Ramachandran R, Schmid SL (2008) Real-time detection reveals that effectors couple dynamin's GTP-dependent conformational changes to the membrane. EMBO J 27(1):27–37
16. Sweitzer SM, Hinshaw JE (1998) Dynamin undergoes a GTP-dependent conformational change causing vesiculation. Cell 93 (6):1021–1029
17. Mears JA, Lackner LL, Fang S, Ingerman E, Nunnari J, Hinshaw JE (2011) Conformational changes in Dnm1 support a contractile mechanism for mitochondrial fission. Nat Struct Mol Biol 18(1):20–26
18. Gasper R, Meyer S, Gotthardt K, Sirajuddin M, Wittinghofer A (2009) It takes two to tango: regulation of G proteins by dimerization. Nat Rev Mol Cell Biol 10(6):423–429
19. Prakash B, Praefcke GJ, Renault L, Wittinghofer A, Herrmann C (2000) Structure of human guanylate-binding protein 1 representing a unique class of GTP-binding proteins. Nature 403(6769):567–571
20. Prakash B, Renault L, Praefcke GJ, Herrmann C, Wittinghofer A (2000) Triphosphate structure of guanylate-binding protein 1 and implications for nucleotide binding and GTPase mechanism. EMBO J 19 (17):4555–4564
21. Chappie JS, Acharya S, Leonard M, Schmid SL, Dyda F (2010) G domain dimerization controls dynamin's assembly-stimulated GTPase activity. Nature 465(7297):435–440
22. Hoppins S, Nunnari J (2009) The molecular mechanism of mitochondrial fusion. Biochim Biophys Acta 1793(1):20–26
23. McNew JA, Sondermann H, Lee T, Stern M, Brandizzi F (2013) GTP-dependent membrane fusion. Annu Rev Cell Dev Biol 29:529–550
24. Cao YL, Meng S, Chen Y, Feng JX, Gu DD, Yu B, Li YJ, Yang JY, Liao S, Chan DC, Gao S (2017) MFN1 structures reveal nucleotide-triggered dimerization critical for mitochondrial fusion. Nature 542(7641):372–376
25. Qi Y, Yan L, Yu C, Guo X, Zhou X, Hu X, Huang X, Rao Z, Lou Z, Hu J (2016) Structures of human mitofusin 1 provide insight into mitochondrial tethering. J Cell Biol 215 (5):621–629
26. Yan L, Qi Y, Huang X, Yu C, Lan L, Guo X, Rao Z, Hu J, Lou Z (2018) Structural basis for GTP hydrolysis and conformational change of MFN1 in mediating membrane fusion. Nat Struct Mol Biol 25(3):233–243
27. Yan L, Sun S, Wang W, Shi J, Hu X, Wang S, Su D, Rao Z, Hu J, Lou Z (2015) Structures of the yeast dynamin-like GTPase Sey1p provide insight into homotypic ER fusion. J Cell Biol 210(6):961–972
28. Bian X, Klemm RW, Liu TY, Zhang M, Sun S, Sui X, Liu X, Rapoport TA, Hu J (2011) Structures of the atlastin GTPase provide insight into homotypic fusion of endoplasmic reticulum membranes. Proc Natl Acad Sci U S A 108 (10):3976–3981
29. Byrnes LJ, Singh A, Szeto K, Benvin NM, O'Donnell JP, Zipfel WR, Sondermann H (2013) Structural basis for conformational switching and GTP loading of the large G protein atlastin. EMBO J 32(3):369–384
30. Byrnes LJ, Sondermann H (2011) Structural basis for the nucleotide-dependent dimerization of the large G protein atlastin-1/SPG3A. Proc Natl Acad Sci U S A 108(6):2216–2221
31. O'Donnell JP, Cooley RB, Kelly CM, Miller K, Andersen OS, Rusinova R, Sondermann H (2017) Timing and reset mechanism of GTP hydrolysis-driven conformational changes of Atlastin. Structure 25(7):997–1010
32. Winsor J, Hackney DD, Lee TH (2017) The crossover conformational shift of the GTPase atlastin provides the energy driving ER fusion. J Cell Biol 216(5):1321–1335
33. Moss TJ, Andreazza C, Verma A, Daga A, McNew JA (2011) Membrane fusion by the GTPase atlastin requires a conserved C-terminal cytoplasmic tail and dimerization through the middle domain. Proc Natl Acad Sci U S A 108(27):11133–11138
34. Liu TY, Bian X, Romano FB, Shemesh T, Rapoport TA, Hu J (2015) Cis and trans interactions between atlastin molecules during membrane fusion. Proc Natl Acad Sci U S A 112(15):E1851–E1860

35. Liu TY, Bian X, Sun S, Hu X, Klemm RW, Prinz WA, Rapoport TA, Hu J (2012) Lipid interaction of the C terminus and association of the transmembrane segments facilitate atlastin-mediated homotypic endoplasmic reticulum fusion. Proc Natl Acad Sci U S A 109(32): E2146–E2154

36. Faust JE, Desai T, Verma A, Ulengin I, Sun TL, Moss TJ, Betancourt-Solis MA, Huang HW, Lee T, McNew JA (2015) The Atlastin C-terminal tail is an amphipathic helix that perturbs the bilayer structure during endoplasmic reticulum homotypic fusion. J Biol Chem 290(8):4772–4783

37. Bigay J, Deterre P, Pfister C, Chabre M (1987) Fluoride complexes of aluminium or beryllium act on G-proteins as reversibly bound analogues of the gamma phosphate of GTP. EMBO J 6(10):2907–2913

38. Hoffman GR, Nassar N, Oswald RE, Cerione RA (1998) Fluoride activation of the Rho family GTP-binding protein Cdc42Hs. J Biol Chem 273(8):4392–4399

39. Aoki K, Kamioka Y, Matsuda M (2013) Fluorescence resonance energy transfer imaging of cell signaling from in vitro to in vivo: basis of biosensor construction, live imaging, and image processing. Develop Growth Differ 55 (4):515–522

40. Garg A, Kaur H, Raghava GP (2005) Real value prediction of solvent accessibility in proteins using multiple sequence alignment and secondary structure. Proteins 61(2):318–324

41. Pugalenthi G, Kandaswamy KK, Chou KC, Vivekanandan S, Kolatkar P (2012) RSARF: prediction of residue solvent accessibility from protein sequence using random forest method. Protein Pept Lett 19(1):50–56

42. Cooley RB, Sondermann H (2017) Probing protein-protein interactions with genetically encoded photoactivatable cross-linkers. Methods Mol Biol 1657:331–345

43. Liu CC, Schultz PG (2010) Adding new chemistries to the genetic code. Annu Rev Biochem 79:413–444

44. Elder AD, Domin A, Kaminski Schierle GS, Lindon C, Pines J, Esposito A, Kaminski CF (2009) A quantitative protocol for dynamic measurements of protein interactions by Förster resonance energy transfer-sensitized fluorescence emission. J R Soc Interface 6:S59–S81

Chapter 9

Analysis of Mitochondrial Membrane Fusion GTPase OPA1 Expressed by the Silkworm Expression System

Tadato Ban and Naotada Ishihara

Abstract

Mitochondria are highly dynamic organelles, which move and fuse to regulate their shape, size, and fundamental function. The dynamin-related GTPases play a critical role in mitochondrial membrane fusion. In vitro reconstitution of membrane fusion using recombinant proteins and model membranes is quite useful in elucidating the molecular mechanisms underlying membrane fusion and to identify the essential elements involved in fusion. However, only a few reconstituting approaches have been reported for mitochondrial fusion machinery due to the difficulty of preparing active recombinant mitochondrial fusion GTPases. Recently, we succeeded in preparing a sufficient amount of recombinant OPA1 involved in mitochondrial inner membrane fusion using a BmNPV bacmid–silkworm expression system. In this chapter, we describe the method for the expression and purification of a membrane-anchored form of OPA1 and liposome-based in vitro reconstitution of membrane fusion.

Key words Mitochondria, Membrane fusion, GTPase protein, Optic atrophy 1(OPA1), In vitro reconstitution, Proteoliposome, Fluorescence resonance energy transfer- (FRET-) based lipid mixing assay, Baculovirus expression system, Silkworm

1 Introduction

Mitochondria play key roles in the regulation of energy production, metabolism, intracellular signaling, and apoptosis [1–3]. In mammalian cells, mitochondria are highly dynamic and change their morphology through frequent fusion and fission. Mitochondria have a double-membrane structure composed of the outer membrane (OM) and inner membrane (IM), and two types of dynamin-related GTPases are required for coordinated OM and IM fusion [4]. In mammals, IM fusion is carried out by IM-localized optic atrophy type 1 (OPA1) [5]. In a steady state, OPA1 exists as the IM-anchored form of OPA1 (L-OPA1) and the transmembrane domain-lacking form of OPA1 (S-OPA1), which is generated from proteolytic processing of L-OPA1. L-OPA1 alone exhibits the significant fusion activity in the cultured cells [6, 7] and

Rajesh Ramachandran (ed.), *Dynamin Superfamily GTPases: Methods and Protocols*, Methods in Molecular Biology, vol. 2159, https://doi.org/10.1007/978-1-0716-0676-6_9,

liposome-based in vitro membrane fusion [8], while the contribution of S-OPA1 in membrane fusion is still controversial [9–12]. OPA1 comprises a GTPase domain with the conserved sequence motif found in other dynamin proteins and predicted self-assembling and mitochondrial-specific phospholipid cardiolipin- (CL-) binding domain downstream of the GTPase domain [2, 8]. However, except the GTPase domain, the identification and the precise contribution of other domains are poorly understood.

The complexity of biological membranes makes it difficult to elucidate the detailed molecular mechanism underlying membrane fusion. Therefore, in vitro reconstitution of membrane fusion using recombinant proteins and model membranes has been used to identify essential elements and understand how these fusion elements mediate membrane fusion [13]. While the reconstitution approaches are insightful in elucidating the molecular mechanism [14, 15], there is a need to prepare the active recombinant protein. Moreover, because of the difficulty of preparing active recombinant mitochondrial fusion GTPases, only a few reconstitution approaches have been reported for mitochondrial fusion machinery.

Several recombinant protein expression systems are currently available. The baculovirus expression system is one of the attractive methods for expressing large amounts of active recombinant proteins, which are difficult to express in bacterial expression. The BmNPV bacmid–silkworm expression system is based on the *Bombyx mori* nucleopolyhedrovirus (BmNPV) and uses silkworm larva as a bioreactor for the expression of recombinant proteins. The silkworm expression system has been used for many human recombinant proteins, including membrane proteins [16]. In addition to the high-expression levels of recombinant proteins compared with those obtained using *Bombyx mori* and Sf9 cell lines [17], silkworm expression has several advantages, such as, low-cost rearing and easy manipulation and scale up mass production of recombinant proteins.

Using silkworm expression and in vitro reconstitution of membrane fusion described in this chapter, we prepared liposome containing human recombinant L-OPA1 and reconstituted the membrane fusion reaction using a well-established fluorescence energy transfer- (FRET-) based lipid mixing assay [14, 18]. From a series of in vitro fusion analyses, we found a unique feature in mitochondrial IM fusion machinery, in which only L-OPA1 on either membrane promotes membrane fusion [8]. Because the silkworm expression system might have advantages for the expression of large membrane proteins, the present method would be applicable to various mitochondrial membrane proteins associated with mitochondrial dynamics.

2 Materials

2.1 Expression and Purification of Recombinant L-OPA1

2.1.1 Preparation of Recombinant BmNPV Bacmid DNA Containing Human L-OPA1 Gene

1. pFastBac vector (Thermo Fisher Scientific).
   pFastBac-HTB-L-OPA1: PCR fragment of human L-OPA1 variant 1 was amplified from Hela cells RNA using the following primers and was subcloned into pFastBac-HTB.
   (a) 5′-GCGGATCCTTTTGGCCAGCAAGATTAG-3′.
   (b) 5′-GCGCGGCCGCTTATTTCTCCTGATGAAGAGC TTC-3′.
2. *E. coli* BmDH10bac (provided by Dr. Katsumi Maenaka, Hokkaido University).
3. LB medium.
4. 10 mg/ml Tetracycline.
5. 10 mg/ml Gentamicin.
6. 50 mg/ml Kanamycin.
7. 100 mM Isopropyl β-D-thiogalactopyranoside (IPTG).
8. 2% 5-Bromo-3-indolyl β-D-galactopyranoside (Bluo-Gal).
9. LB agar plate containing 50 μg/ml kanamycin and 7 μg/ml gentamycin.
10. Plasmid purification kit.
11. Universal primers:
    (a) M13 forward (−40) 5′-GTTTTCCCAGTCACGAC-3′
    (b) M13 reverse 5′-CAGGAAACAGCTATGAC-3′.
12. DNA polymerase for PCR and related reagents.
13. 0.8% (wt/vol) agarose gel and DNA electrophoresis-related reagents.

2.1.2 Expression of Recombinant Human L-OPA1

1. DMRIE-C (Thermo Fisher Scientific).
2. Fifth-instar silkworm (*Bombyx mori*) larva.
3. Artificial diet silkmate 2S (Nihon Nosan Kogyo).
4. Climate chamber.
5. Glass–Teflon potter homogenizer (50 ml).

2.1.3 Purification of Recombinant L-OPA1 from Silkworm Fat Body

1. PB buffer: 50 mM NaPi (pH 8.0), 500 mM NaCl, 10% glycerol, 1 mM DTT, 1 mM PMSF, and 0.5% sodium thiosulfate.
2. Centrifuge.
3. Dodecyl maltoside (DDM).
4. Ni-chelating beads.
5. 2 M imidazole in PB buffer.
6. Tube rotator.

7. RB buffer: 50 mM Tris–HCl (pH 7.4), 300 mM NaCl, 10% glycerol, and 1 mM DTT.
8. 2 M Imidazole in RB buffer.
9. 1.0 × 5 cm Chromatography column.
10. 7.5% (wt/vol) polyacrylamide gel and SDS-PAGE-related reagents.
11. Bradford reagent.
12. Spectrophotometer.

2.2 Preparation of L-OPA1 Proteoliposome

2.2.1 Detergent Exchange from DDM to MEGA 8

1. MEGA 8.
2. Ni-chelating beads.
3. 1.0 × 5 cm Chromatography column.

2.2.2 Preparation of L-OPA1 Proteoliposome

1. Lipids from Avanti Polar Lipids. Lipid stocks are dissolved in chloroform and stored at −30 °C.
   (a) 50 mg/ml 1-Palmitoyl-2-oleoyl-*sn*-glycero-3-phosphocholine (POPC).
   (b) 10 mg/ml 1-Palmitoyl-2-oleoyl-*sn*-glycero-3-phosphoethanolamine (POPE).
   (c) 10 mg/ml L-α-Phosphatidylinositol (soy-PI).
   (d) 10 mg/ml 1′,3′-bis[1,2-Dioleoyl-*sn*-glycero-3-phospho]-*sn*-glycerol (CL$(18{:}1)_4$).
2. Fluorescent-labeled lipids from Molecular Probes. Lipid stocks are dissolved in chloroform and stored at −30 °C.
   (a) 5 mg/ml (*N*-(7-Nitrobenz-2-oxa-1,3-diazol-4-yl)-1,2-dihexadecanoyl-*sn*-glycero-3-phosphoethanolamine (NBD-PE).
   (b) 2.5 mg/ml Lissamine rhodamine B 1,2-dihexadecanoyl-*sn*-glycero-3-phosphoethanolamine (Rh-PE).
3. Glass tube.
4. 10 and 50 μl Glass syringes.
5. N_2 gas tank.
6. Vacuum pump or house vacuum line.
7. Tube rotator.
8. 200 μl PCR tube.
9. 100 μl Dialysis button (Hampton Research).
10. MWCO of 30 kDa dialysis membrane.

2.3 In Vitro Membrane Fusion Assay

1. 384-Well polystyrene black microplate.
2. Fluorescence microplate reader (λ excitation = 465 nm, λ emission = 540 nm).
3. 50 mM GTP and 50 mM $MgCl_2$ in RB buffer.
4. Multichannel pipette.
5. 4% Triton X-100 in RB buffer.

3 Method

3.1 Expression and Purification of Recombinant L-OPA1

In the present baculovirus expression system, baculovirus shuttle vector (bacmid) and "Bac to Bac" technology are employed for the high-efficient and rapid production of the recombinant baculovirus (Fig. 1a). Details of BmNPV bacmid–silkworm expression system are described in previous reports [16, 17].

3.1.1 Preparation of Recombinant BmNPV Bacmid DNA Containing Human L-OPA1 Gene

1. Mix 100–1000 ng of pFastBac-HTB-L-OPA1 and 100 μl of BmDH10bac competent cells on ice (*see* **Note 1**).
2. Heat shock at 42 °C for 45 s and place on ice for 2 min.
3. Add 1.4 ml of LB medium and incubate at 37 °C for 1 h with shaking at 180 rpm.
4. Add 1.4 μl of tetracycline and continue incubation overnight.
5. Add 1 μl of gentamicin and continue incubation for 2 h.
6. Spread 20 μl of IPTG and 70 μl of Bluo-Gal on an LB agar plate containing kanamycin and gentamycin (*see* **Note 2**).
7. Plate 150 μl of the cultured BmDH10bac, and incubate at 37 °C overnight (*see* **Note 3**).
8. Pick white colonies, inoculate into 4 ml of LB medium containing 50 μg/ml kanamycin and incubate at 37 °C overnight with shaking at 180 rpm.
9. Harvest 3 ml of bacterial cells by centrifuging at 4000× *g* for 10 min at 4 °C and isolate bacmid DNA using a plasmid purification kit. Keep the remaining bacterial cells at 4 °C to make a glycerol stock.
10. Confirm the correct transposition in bacmid DNA using PCR with the universal primers, M13 forward (−40) and M13 reverse (*see* **Note 4**).
11. Analyze the PCR products using an agarose gel electrophoresis. In case of L-OPA1, if transposition has occurred, you can see the band at 5000 bp.
12. Make glycerol stocks of BmDH10bac with the correct transposition. Mix 750 μl of the culture with 150 μl of sterile 60% glycerol in a 1.5 ml tube, then flash-freeze in liquid nitrogen and store at −80 °C.

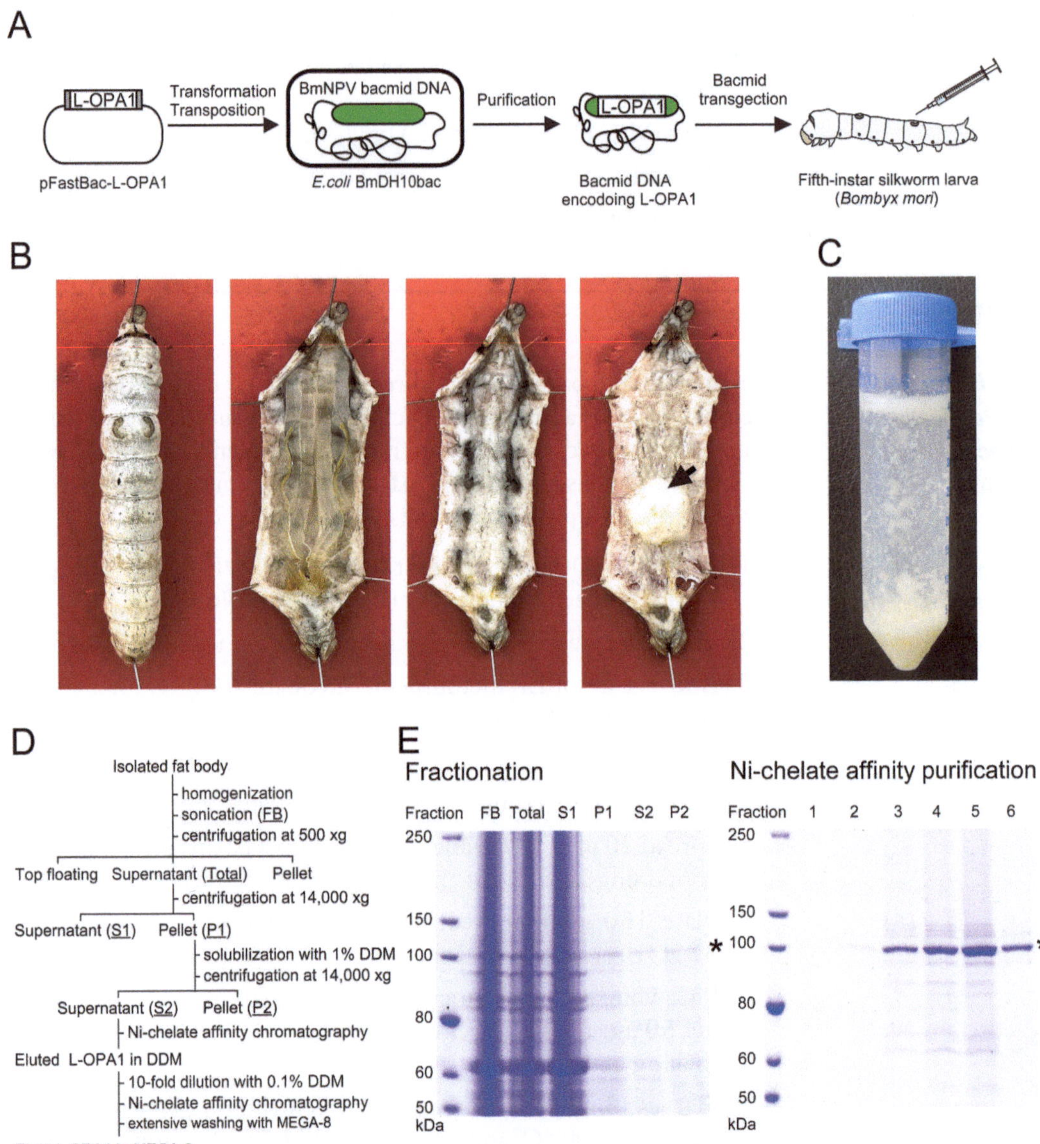

Fig. 1 Expression and purification of recombinant human L-OPA1 using silkworm larvae. (**a**) Schematic diagram of the BmNPV bacmid–silkworm expression. (**b**) Dissection of silkworm larva. The fat bodies expressing L-OPA1 were collected manually. Arrow indicates collected fat body. (**c**) The isolated fat bodies from 15 silkworm larvae. (**d**) Procedures for the fractionation and purification of recombinant L-OPA1 from fat bodies. L-OPA1 was solubilized by DDM and purified by Ni-chelate affinity chromatography. (**e**) SDS-PAGE analysis and staining with Coomassie Blue of fractionated and purified L-OPA1. Asterisk indicates purified L-OPA1

13. Streak the small piece of frozen glycerol stock onto an LB agar plate containing 50 μg/ml kanamycin and incubate overnight at 37 °C.

14. Pick a single colony, inoculate into 3 ml of LB medium containing 50 μg/ml kanamycin, and incubate at 37 °C overnight with shaking at 180 rpm.
15. Transfer the 1.5 ml overnight culture into 100 ml LB medium containing 50 μg/ml kanamycin and incubate at 37 °C overnight with shaking at 180 rpm.
16. Harvest bacterial cells by centrifuging at 4000× *g* for 10 min at 4 °C (*see* **Note 5**).
17. Isolate bacmid DNA using a plasmid purification kit, and store at 4 °C (*see* **Note 6**). The recombinant bacmid DNA should be used within a week.

3.1.2 Expression of Recombinant Human L-OPA1 in Silkworm Fat Body

To generate the recombinant BmNPV encoding L-OPA1 gene, the purified bacmid DNA is directly injected into silkworm larva by lipofection. BmNPV amplifies in hemocytes and fat body of silkworm larvae, and L-OPA1 is expressed and localized in the fat body. Therefore, isolation of infected fat body is required before the purification.

1. Mix 1.5 μg of bacmid DNA with 3 μl of DMRIE-C per silkworm larva, incubate for 45 min at room temperature, and dilute with 50 μl of sterile ultrapure water. Typically, 60 silkworm larvae are injected at the same time. Therefore, mix 1.5 μg × 60 = 90 μg bacmid DNA with 3 μl × 60 = 180 μl of DMRIE-C. After incubation, add 50 μl × 60 = 3000 μl of Ultra Pure Water.
2. Inject 50 μl of bacmid–DMRIE-C mixture into the dorsal side of each silkworm larva (first day of the fifth-instar larval stage) using a 30 gauge needle (Fig. 1a).
3. After injection, pause for a few seconds before pulling out the needle to avoid leakage of the bacmid mixture.
4. Transfer the larvae to a rearing box with breathing holes. Rear for 6 days in a climate chamber maintained at 25 °C and 80% humidity with regular feeding. We generally rear 15 silkworm larvae in a 22 × 16 × 6.5 cm box.
5. The 15 larvae are fed on an artificial diet every morning as follows: day 1 (20 g), day 2 (30 g), day 3 (45 g), day 4 (50 g), day 5 (50 g), and day 6 (35 g).
6. Dissect silkworm larvae and isolate infected fat bodies (Fig. 1b). Fat bodies are stored from 15 silkworm larvae in a 50 ml conical tube containing 40 ml of ice-cold PB buffer (Fig. 1c).
7. Homogenize the collected fat bodies using a 50 ml Glass–Teflon potter homogenizer with 10 strokes at 1300 arm, then flash-freeze in liquid nitrogen and store at −80 °C.

3.1.3 Purification of Recombinant Human L-OPA1 from Silkworm Fat Body

The procedure described in this section is for L-OPA1 purification from fat bodies isolated from 60 silkworm larvae. As described in Subheading 3.1.2, the fat bodies isolated from 15 silkworms are stored in one 50 ml conical tube. Therefore, in the subsequent procedures (**steps 1–15**), four tubes are handled simultaneously, except in the sonication step (**step 2**). The silkworm-expressed recombinant L-OPA1 has an additional 6× His sequence at the N-terminus. After the solubilization with DDM, the recombinant L-OPA1 is purified using a Ni-chelate affinity chromatography.

1. Thaw the frozen fat bodies on ice.
2. Transfer the fat bodies into a 50 ml beaker and sonicate the fat bodies on ice (FB in Fig. 1d) (*see* **Note 7**).
3. Transfer the homogenized fat bodies into a new 50 ml conical tube and centrifuge at 500 × *g* for 5 min at 4 °C.
4. Decant the supernatant (Total in Fig. 1d) into a new 50 ml conical tube carefully to avoid mixing of the top floating and large debris at the bottom of the tube.
5. Centrifuge the supernatant at 14,000 × *g* for 1 h at 4 °C.
6. After discarding the top floating using a spatula, remove the supernatant (S1 in Fig. 1d) carefully using a pipette.
7. Resuspend the pellet (P1 in Fig. 1d) in 2 ml of ice-cold PB buffer containing 10 mM imidazole and 1% DDM and transfer to a new 50 ml conical tube.
8. Add 38 ml of ice-cold PB buffer containing 10 mM imidazole and 1% DDM and incubate for 2 h at 4 °C with gentle agitation.
9. Centrifuge the solubilized pellet at 14,000× *g* for 30 min at 4 °C and collect the supernatant (S2 in Fig. 1d).
10. Equilibrate 750 μl (bed volume) of Ni-chelating beads by washing three times with 1 ml pure water and PB buffer containing 10 mM imidazole and 1% DDM at **step 9** (*see* **Note 8**).
11. Incubate the supernatant with Ni-chelating beads overnight at 4 °C with gentle agitation.
12. Centrifuge at 50 × *g* for 1 min at 4 °C and remove the supernatant.
13. Add 1 ml of PB buffer containing 10 mM imidazole and 1% DDM, centrifuge at 50 × *g* for 1 min at 4 °C, and remove the supernatant.
14. Repeat **step 13** two more times.
15. Resuspend in 1 ml of ice-cold PB buffer containing 10 mM imidazole and 1% DDM, apply the beads into a chromatography column from the four 50 ml conical tubes carefully.

16. Wash the beads with 15 ml (5 column volumes) of ice-cold RB buffer containing 10 mM imidazole and 0.1% DDM.
17. Wash the beads with 15 ml (5 column volumes) of ice-cold RB buffer containing 20 mM imidazole and 0.1% DDM.
18. Elute the L-OPA1 with 1.5 ml of ice-cold RB buffer containing 250 mM imidazole and 0.1% DDM.
19. Repeat **step 18** seven more times.
20. Analyze the eluted fractions using SDS-PAGE, and determine the concentration using a Bradford assay. In our case, the total yield was more than 2 mg from 60 silkworm larvae.
21. Pool the peak fractions, flash–freeze in liquid nitrogen and store at −80 °C.

3.2 Preparation of L-OPA1 Proteoliposome

3.2.1 Detergent Exchange from DDM to MEGA 8

To analyze L-OPA1 function in membrane fusion, in vitro membrane fusion was reconstituted using L-OPA1 proteoliposomes. Purified L-OPA1 was inserted into the mitochondrial IM-mimicked liposomes by detergent-dialysis method (Fig. 2a) [19, 20]. Because of its low critical micellar concentration (CMC) detergent, DDM is not preferred for detergent-dialysis method. Therefore, DDM was replaced with the high CMC detergent, MEGA 8, before the preparation of L-OPA1 proteoliposomes.

1. Equilibrate 1 ml (bed volume) of Ni-chelating beads by washing three times with 1 ml of pure water and RB buffer containing 10 mM imidazole and 0.1% DDM, and pour into a chromatography column.
2. Thaw the frozen purified L-OPA1 from 60 silkworm larvae on ice.
3. Dilute the purified L-OPA1 ten times with ice-cold RB buffer containing 10 mM imidazole and 0.1% DDM, and apply into a column packed with the Ni-chelating beads by gravity flow at 4 °C.
4. Wash the beads with 20 ml (20 column volumes) of ice-cold RB buffer containing 2.5% MEGA 8.
5. Elute the L-OPA1 with 0.5 ml of ice-cold RB buffer containing 250 mM imidazole and 2.5% MEGA 8.
6. Repeat **step 5** seven more times.
7. Analyze the eluted fractions using SDS-PAGE and determine the concentration using a Bradford assay. In our case, recovery rate was about ~70% after the exchange from DDM to MEGA 8.
8. Aliquot the peak fractions into 100 μl aliquots, flash–freeze in liquid nitrogen, and store at −80 °C.

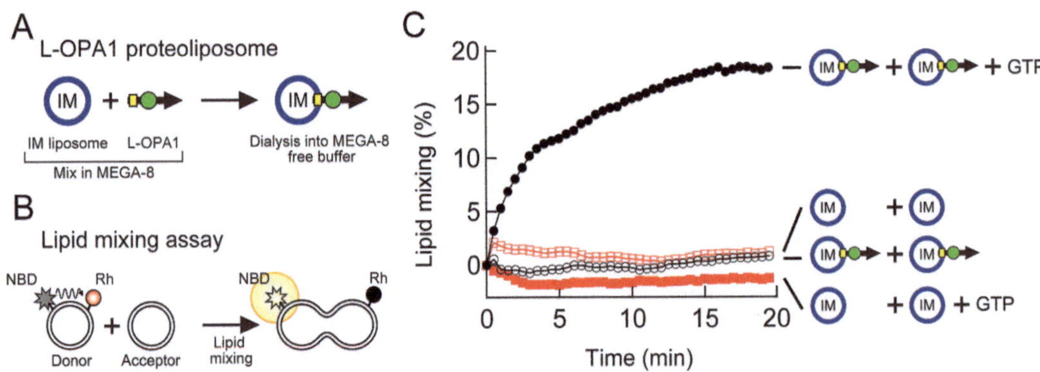

Fig. 2 Preparation of proteoliposomes containing L-OPA1 and in vitro membrane fusion. (**a**) Recombinant L-OPA1 and IM liposomes were mixed in the presence of a detergent, and proteoliposomes were reconstituted by dialysis. IM liposome means mitochondrial IM- mimicked liposome (42% POPC, 25% POPE, 8% soy-PI, and 25% CL). For fluorescence-labeled donor liposome, 3% of POPE was replaced for fluorescence-labeled lipids. (**b**) Schematic diagram of FRET-based lipid mixing assay. (**c**) L-OPA1-mediated membrane fusion in vitro

3.2.2 Preparation of L-OPA1 Proteoliposome

In vitro membrane fusion reaction was analyzed using the well-established fluorescence resonance energy transfer- (FRET-) based lipid mixing assay. The fluorescent 7-nitrobenz-2-oxa-1,3-diazol-4-yl (NBD) and rhodamine- (Rh-) labeled donor liposome and nonfluorescent-labeled acceptor liposome are used in the lipid mixing assay. When the liposomes are fused, fluorescent lipids are diffused in the resulting membrane, and the NBD fluorescence can be detected previously quenched by Rh (Fig. 2b). Therefore, both donor L-OPA1 proteoliposome and acceptor L-OPA1 proteoliposome are required for the analysis. The lipid compositions of liposome are designed to mimic the lipid composition of IM [21]. We describe the method for preparing the L-OPA1 proteoliposomes contacting L-OPA1 and IM lipid at the molar ratio of 1:1000 (2 μM L-OPA1 and 2 mM IM lipids) (*see* **Note 9**).

1. Prepare 100 μl of 8 mM IM lipid mixtures. The lipid compositions are as follows (percent molar): "fluorescent-labeled donor IM" liposome (POPC:POPE:soy-PI:CL$(18{:}1)_4$:NBD–PE:Rho–PE = 42:22:8:25:1.5:1.5), "nonfluorescent-labeled acceptor IM" liposome (POPC:POPE:soy-PI:CL $(18{:}1)_4$ = 42:25:8:25). Transfer the desired volume of stock lipid solution in chloroform into a small glass tube.
2. Remove chloroform under a nitrogen gas (N_2) stream until the lipids form a film on the side of glass tube.
3. Dry the lipid film using a vacuum pump for 30 min to remove any trace chloroform.
4. Add 100 μl of RB buffer containing 5% MEGA 8 to a lipid concentration of 8 mM, and vortex gently until the solution becomes clear.

5. Mix 30 μl of lipid mixture (8 mM), 80 μl of purified L-OPA1 (3 μM), and 10 μl of RB buffer containing 2.5% MEGA 8 in a 200 μl PCR tube (L-OPA1 to lipid molar ratio = 1:1000).
6. Incubate for 2 h at 4 °C with gentle agitation.
7. Transfer 100 μl of detergent–protein–lipid mixtures into a 100 μl dialysis button with MWCO of 30 kDa dialysis membrane.
8. Put dialysis button into a 50 ml conical tube containing 50 ml RB buffer and dialyze overnight at 4 °C with gentle agitation.
9. Change dialysis buffer and continue dialysis for 8 h.
10. Change dialysis buffer and continue dialysis overnight.
11. Harvest proteoliposomes from the dialysis button and transfer into a 1.5 ml tube.
12. Store proteoliposomes at 4 °C. These should be used within a week.

3.3 *In Vitro* Membrane Fusion

As described in Subheading 3.2.2, in vitro membrane fusion reaction was carried out by lipid mixing assay. The lipid mixing assay was performed in a fluorescence microplate reader. In the presence of GTP, efficient membrane fusion was observed in L-OPA1 proteoliposomes (Fig. 2c)

1. Transfer 11 μl of RB buffer into each well of a black 384-well plate.
2. Add 5 μl of nonfluorescent-labeled acceptor L-OPA1 proteoliposomes and 2 μl of fluorescent-labeled donor L-OPA1 proteoliposomes.
3. After mixing with gentle pipetting, incubate for 10 min at 30 °C in a preheated fluorescence microplate reader (*see* **Note 10**).
4. Add 2 μl of the GTP–$MgCl_2$ mixtures using a multichannel pipette and immediately start the measurement (*see* **Note 11**).
5. Measure NBD fluorescence at 540 nm with an excitation of 465 nm at 30 s intervals at 30 °C.
6. Add 2 μl of Triton X-100 to each sample to terminate the reaction.
7. After mixing the sample well by pipetting, incubate for 10 min at 30 °C.
8. Measure NBD fluorescence and obtain maximum NBD fluorescence.
9. Normalize the measured NBD fluorescence signal obtained from each reaction by the following equation [14, 18]:

   Lipid mixing (%) = $(F_t - F_0)/(F_{max} - F_0) \times 100$, in which F_t is the NBD fluorescence during the measurement, F_0 is the initial NBD fluorescence immediately after the addition of GTP–$MgCl_2$, and F_{max} is the NBD fluorescence after the addition of Triton X-100 to the reaction mixture.

4 Notes

1. BmDH10bac provided by Dr. Maenaka, Hokkaido University, not the commercial BmDH10, is needed for a BmNPV bacmid-silkworm expression system.
2. Bluo-Gal is recommended for blue/white color selection because Bluo-Gal produces a darker blue color than X-gal.
3. For the first time, serially dilute the cultured BmDH10bac with LB medium and plate on agar plate to have well-separated colonies.
4. We typically use KOD DNA polymerase (Toyobo) for PCR, but other DNA polymerases are also possible.
5. Bacterial cell pellets can be stored at −30 °C.
6. We use Qiafilter Plasmid Maxi Kit (Qiagen) for 100 ml of cultured cells and obtain more than 200 μg of recombinant bacmid DNA. Other DNA purification kits may also be used.
7. Ultrasonic intensity depends on equipment used. In our case, sonication is performed using an Astrason model W-385, equipped with a flat tip. Ultrasonic duration is 10 min at an ultrasonic cycle time 2, duty cycle 50, and output control 9.
8. We use Ni Sepharose® 6 Fast Flow (GE Healthcare), but other Ni-chelating beads may also be used.
9. We previously showed that significant membrane fusion is detected in the proteoliposome with L-OPA1 to lipid molar ratio greater than 1:4000 (0.5 μM L-OPA1 and 2 mM IM lipids) [9].
10. Lipid mixing assay is performed at 30 °C to avoid evaporation of the samples.
11. Multichannel pipette facilitates synchronization of the start of membrane fusion reaction.

Acknowledgments

We thank Dr. T. Oka (Rikkyo University) and Dr. K. Maenaka (Hokkaido University) for advice on the silkworm expression system and Dr. J. Mima (Osaka University) for advice on the preparation of assay with proteoliposomes. This work is supported by JSPS KAKENHI grant number 18K06096, MEXT-Supported Program for the Strategic Research Foundation at Private Universities, the

Takeda Science Foundation (T.B.), the Naito Foundation (T.B.) and the Ichiro Kanehara Foundation (T.B.).

References

1. Ishihara N, Otera H, Oka T, Mihara K (2013) Regulation and physiologic functions of GTPases in mitochondrial fusion and fission in mammals. Antioxid Redox Signal 19:389–399
2. Labbe K, Murley A, Nunnari J (2014) Determinants and functions of mitochondrial behavior. Annu Rev Cell Dev Biol 30:357–391
3. Mishra P, Chan DC (2016) Metabolic regulation of mitochondrial dynamics. J Cell Biol 212:379–387
4. McNew JA, Sondermann H, Lee T, Stern M, Brandizzi F (2013) GTP-dependent membrane fusion. Annu Rev Cell Dev Biol 29:529–550
5. MacVicar T, Langer T (2016) OPA1 processing in cell death and disease—the long and short of it. J Cell Sci 129:2297–2306
6. Ishihara N, Fujita Y, Oka T, Mihara K (2006) Regulation of mitochondrial morphology through proteolytic cleavage of OPA1. EMBO J 25:2966–2977
7. Tondera D, Grandemange S, Jourdain A, Karbowski M, Mattenberger Y, Herzig S, Da Cruz S, Clerc P, Raschke I, Merkwirth C, Ehses S, Krause F, Chan DC, Alexander C, Bauer C, Youle R, Langer T, Martinou JC (2009) SLP-2 is required for stress-induced mitochondrial hyperfusion. EMBO J 28:1589–1600
8. Ban T, Ishihara T, Kohno H, Saita S, Ichimura A, Maenaka K, Oka T, Mihara K, Ishihara N (2017) Molecular basis of selective mitochondrial fusion by heterotypic action between OPA1 and cardiolipin. Nat Cell Biol 19:856–863
9. Song Z, Chen H, Fiket M, Alexander C, Chan DC (2007) OPA1 processing controls mitochondrial fusion and is regulated by mRNA splicing, membrane potential, and Yme1L. J Cell Biol 178:749–755
10. Anand R, Wai T, Baker MJ, Kladt N, Schauss AC, Rugarli E, Langer T (2014) The i-AAA protease YME1L and OMA1 cleave OPA1 to balance mitochondrial fusion and fission. J Cell Biol 204:919–929
11. Ban T, Heymann JA, Song Z, Hinshaw JE, Chan DC (2010) OPA1 disease alleles causing dominant optic atrophy have defects in cardiolipin-stimulated GTP hydrolysis and membrane tubulation. Hum Mol Genet 19:2113–2122
12. Ban T, Kohno H, Ishihara T, Ishihara N (2018) Relationship between OPA1 and cardiolipin in mitochondrial inner-membrane fusion. Biochim Biophys Acta 1859:951–957
13. Wickner W, Schekman R (2008) Membrane fusion. Nat Struct Mol Biol 15:658–664
14. Weber T, Zemelman BV, McNew JA, Westermann B, Gmachl M, Parlati F, Sollner TH, Rothman JE (1998) SNAREpins: minimal machinery for membrane fusion. Cell 92:759–772
15. Orso G, Pendin D, Liu S, Tosetto J, Moss TJ, Faust JE, Micaroni M, Egorova A, Martinuzzi A, McNew JA, Daga A (2009) Homotypic fusion of ER membranes requires the dynamin-like GTPase atlastin. Nature 460:978–983
16. Kato T, Kajikawa M, Maenaka K, Park EY (2010) Silkworm expression system as a platform technology in life science. Appl Microbiol Biotechnol 85:459–470
17. Kajikawa M, Sasaki-Tabata K, Fukuhara H, Horiuchi M, Okabe Y, Maenaka K (2012) Silkworm baculovirus expression system for molecular medicine. J Biotechnol Biomaterial S9:005
18. Scott BL, Van Komen JS, Liu S, Weber T, Melia TJ, McNew JA (2003) Liposome fusion assay to monitor intracellular membrane fusion machines. Methods Enzymol 372:274–300
19. Rigaud JL, Levy D (2003) Reconstitution of membrane proteins into liposomes. Methods Enzymol 372:65–86
20. Mima J, Wickner W (2009) Complex lipid requirements for SNARE- and SNARE chaperone-dependent membrane fusion. J Biol Chem 284:27114–27222
21. Ardail D, Privat JP, Egret-Charlier M, Levrat C, Lerme F, Louisot P (1990) Mitochondrial contact sites. Lipid composition and dynamics. J Biol Chem 265:18797–18802

Chapter 10

Cell-Free Analysis of Mitochondrial Fusion by Fluorescence Microscopy

Nyssa Becker Samanas and Suzanne Hoppins

Abstract

Dynamin-related proteins on both the mitochondrial outer and inner membranes mediate membrane fusion. Mitochondrial fusion is regulated in many different physiological contexts including cell cycle progression, differentiation pathways, stress responses, and cell death. Mitochondrial fusion is opposed by mitochondrial division and requires movement of mitochondria on microtubules. We developed a cell-free reconstituted mitochondrial fusion assay to circumvent the complexity of the pathways impinging on the activity of the mitochondrial fusion machinery in vivo. This allows for quantification of mitochondrial fusion in defined conditions and in the absence of other processes such as mitochondrial division or transport. The impact of proteins or small molecules on mitochondria fusion can also be assessed. Here we describe the cell-free mitochondrial fusion assay using mitochondria isolated from mouse embryonic fibroblasts.

Key words Mitochondrial dynamics, Fusion, Cell-free, Isolated mitochondria, Mitofusin, Opa1, Microscopy

1 Introduction

The state of the mitochondrial network is both a contributor to and indicator of cellular health and function. The overall structure of the network is modulated by the opposing processes of mitochondrial fusion and division, which are both directly controlled by dynamin-related proteins. In mammals, dynamin-related protein 1 (Drp1) is responsible for mitochondrial division, while mitofusin 1 and mitofusin 2 (Mfn1 and Mfn2, respectively, Mfns, collectively) and optic atrophy 1 (Opa1) mediate outer and inner mitochondrial membrane fusion, respectively [1–3]. The importance of these processes is highlighted by the finding that loss of any component is embryonic lethal in mouse models [4–8]. Additionally, single-allele mutant variants of Drp1, Mfn2, and Opa1 are associated with neuropathies [9–12]. Therefore, dissecting the mechanisms of mitochondrial dynamics and probing their regulatory pathways

Rajesh Ramachandran (ed.), *Dynamin Superfamily GTPases: Methods and Protocols*, Methods in Molecular Biology, vol. 2159, https://doi.org/10.1007/978-1-0716-0676-6_10,

are important to understand the fundamental physiology of the cell and the progression of various diseases.

Live- or fixed-cell imaging of the mitochondrial network using fluorescent markers is widely utilized to query mitochondrial dynamics. While imaging mitochondria in intact cells informs about the overall state of the mitochondrial network under different conditions, this approach has limitations. Live- or fixed-cell imaging is generally a qualitative description of the network state rather than a quantitative measure of either mitochondrial fusion or division and does not inform on either of these processes individually or over time. Such limitations have been addressed using methods such as fluorescence recovery after photobleaching (FRAP) and photoactivatable fluorescent protein analyses, as recently reviewed in [13]. These methods allow assessment of both the degree of connectivity in the mitochondrial network and network dynamics over time in situ. However, analyzing the state of the network as a whole can be complicated by changes in numerous regulatory pathways that impinge on both the division and fusion proteins, thus limiting the ability to draw conclusions solely about a single protein or pathway of interest. Here, we describe in detail a method to quantify mitochondrial fusion in a cell-free assay as first presented in [14]. As a reconstituted system, this method measures only mitochondrial fusion and, therefore, is a specific inquiry into the function of the mitochondrial fusion dynamin-related proteins and the proteins that regulate their activity.

In this assay, mitochondria are isolated from two populations of cells stably expressing fluorescent mitochondrial matrix markers of different colors. Isolated, differentially labeled mitochondria are mixed, concentrated by centrifugation, and resuspended in buffer that supports mitochondrial outer and inner membrane fusion. Following visualization by confocal or widefield microscopy, fusion is quantified by comparing the number of fused organelles, the mitochondria with both matrix markers, to the total number of mitochondria. In its simplest form, with only two mitochondrial populations and fusion buffer, the basal rate of mitochondrial fusion from the cells is measured, which allows a direct readout of the activity of the dynamin-related fusion proteins. The assay can then be further exploited to determine the role of other factors including pure protein, crude cytosol, or small molecules.

2 Materials

All solutions should be prepared using ultrapure water and analytical grade reagents. Filter sterilization should be performed with 0.2 μm filters. Dispose of all reagents according to local guidelines.

1. Media appropriate to the cells being used. Complete media for all cells specified in this protocol is: DMEM + high glucose and GlutaMAX™ (Gibco/Fisher Scientific) supplemented with 10% fetal bovine serum and 1% penicillin/streptomycin.
2. Plasmids encoding two different mitochondrial matrix-targeted fluorescent proteins, such as Addgene #58425 (pclbw-mitoTagRFP) and 58,426 (pclbw-mitoCFP) [15].
3. Retroviral packaging cell line Platinum-E (Plat-E) cells (Cell Biolabs, Inc. #RV-101).
4. Mouse embryonic fibroblasts or cells of your choice.
5. Transfection media: We use Opti-MEM™ (Gibco #31985–062).
6. FuGENE™ HD transfection reagent (Promega #E2311).
7. Puromycin and blasticidin antibiotics.
8. Sterile luer lock syringes.
9. 0.45 μm PES membrane syringe filters.
10. Polybrene infection reagent.
11. Freezing media: DMEM + high glucose (no glutamine) supplemented with 20% FBS and 10% tissue culture quality DMSO.
12. Coverslips or glass bottom dishes (such as MatTek #P35G-1.5-20-C).
13. 0.1 M Tris-MOPS pH 7.4: Dissolve 1.21 g Tris powder in about 50 mL water. Adjust the pH of the solution to 7.4 with MOPS powder and bring volume to 100 mL. Filter sterilize and store at 4 °C.
14. 1 M sucrose: Dissolve 85.58 g sucrose in water to a total volume of 250 mL. Filter sterilize and store at 4 °C.
15. 0.1 M EGTA-Tris: Dissolve 3.8 g EGTA in about 50 mL water. Adjust the pH of the solution to 7.4 with Tris powder and bring the final volume to 100 mL. Filter sterilize and store at room temperature.
16. 3 M sorbitol: Dissolve 136.5 g sorbitol in water to a final volume of 250 mL. Filter sterilize and store at 4 °C.
17. 0.5 M PIPES pH 6.8: Dissolve 15.12 g PIPES in about 50 mL water. Adjust the pH of the solution to 6.8 with KOH and bring the final volume to 100 mL (*see* **Note 1**). Filter sterilize and store at room temperature.
18. 1 M magnesium acetate: Dissolve 10.72 g magnesium acetate tetrahydrate in water to a final volume of 50 mL. Filter sterilize and store at room temperature.

19. 5 M potassium acetate: Dissolve 24.54 g potassium acetate in water to a final volume of 50 mL. Filter sterilize and store at room temperature.
20. Mitochondrial isolation buffer (MIB): 0.01 M Tris-MOPS (pH 7.4), 0.2 M sucrose, 1 mM EGTA-Tris. Make fresh each day and keep on ice or at 4 °C.
21. Cytosol buffer: 20 mM PIPES pH 6.8, 150 mM potassium acetate, 5 mM magnesium acetate, 0.4 M sorbitol. Make fresh each day and store on ice.
22. 30 mg/mL creatine kinase: Resuspend creatine kinase powder in cytosol buffer at working concentration of 30 mg/mL (*see* **Note 2**).
23. 1 M creatine phosphate: Dissolve 211.11 mg creatine phosphate in 1 mL cytosol buffer. Store 10 μL aliquots at −80 °C.
24. 250 mM GTP: Dissolve 25 mg GTP in 191 μL cytosol buffer. Adjust the pH of the solution to 7.0 with 2 M NaOH (*see* **Note 3**). Store 5 μL aliquots at −80 °C. On the day of the assay, dilute one aliquot to 25 mM with cytosol buffer.
25. 250 mM ATP: Dissolve 25 mg ATP in 197 μL cytosol buffer. Adjust the pH of the solution to 7.0 with 2 M NaOH (*see* **Note 3**). Store 5 μL aliquots at −80 °C. On the day of the assay, dilute one aliquot to 25 mM with cytosol buffer.
26. Fusion buffer: 1.2 mg/mL creatine kinase, 40 mM creatine phosphate, 1.5 mM ATP, 1.5 mM GTP in cytosol buffer.
27. 3% low melt agarose: Dissolve 30 mg low melt agarose in 1 mL cytosol buffer by warming the solution to 90 °C. Keep at 65 °C.
28. Kontes Potter-Elvehjem tissue grinder, 4 mL (such as Fisher Scientific #K885510-0020).
29. Motor-driven homogenizer (such as Caframo #BDC3030).
30. Single-well glass slides and covers (such as Fisher Scientific #S175201).
31. Clear nail polish.
32. Fluorescence widefield or confocal microscope.

3 Method

3.1 Expressing Fluorescent Mitochondrial Matrix-Targeted Proteins in Mouse Embryonic Fibroblasts (MEFs) (Fig. 1)

For all experiments, cells are maintained at 37 °C and 5% CO_2.

1. Day 1: Split Plat-E cells, which have been maintained in complete media supplemented with 1 μg/mL puromycin and 10 μg/mL blasticidin, and plate at approximately 80% confluency (approximately 3.5 × 10^5 cells in each well of a six-well dish) in 2.5 mL complete media (no antibiotics) (*see* **Note 4**).

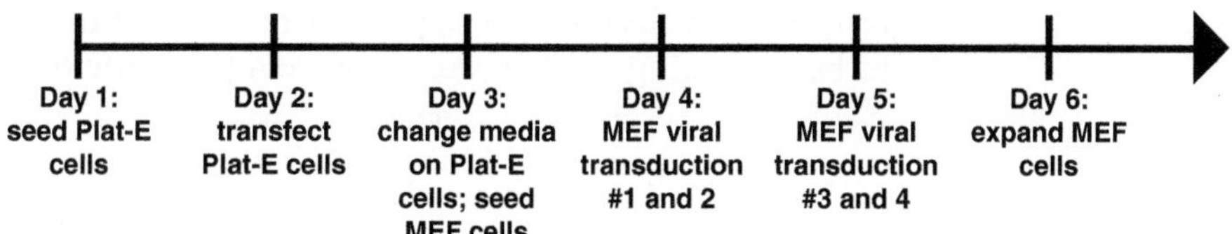

Fig. 1 Timeline of generating stable cell lines expressing fluorescent mitochondrial matrix-targeted proteins by retroviral production and infection

2. Grow Plat-E cells overnight.
3. Day 2: Transfect Plat-E cells with high purity, low endotoxin plasmid DNA (mitochondria-targeted RFP and CFP seperately). Combine 175 μL Opti-MEM with 3 μg of DNA and mix. Add 10 μL of FuGENE™ HD directly to the liquid and mix vigorously. Incubate at room temperature for 15 min before dropping the mixture onto the Plat-E cells.
4. Incubate Plat-E cells with transfection reagents for 8–24 h.
5. Day 3: Aspirate all media from Plat-E cells and replace with 2.5 mL complete media. Incubate for ≥24 h. At the end of this incubation, you will collect the first virus-containing media for infection.
6. Day 3: 10–24 h before first infection, plate MEF cells (or cell of your choice) at about 70% confluency (approximately 1×10^5 cells per well of six-well dish) (*see* **Note 4**).
7. Day 4: Approximately 48 h after the transfection of the Plat-E cells, perform first viral infection. Aspirate media from MEF cells and discard. Collect media from Plat-E cells with a sterile syringe. Pass this virus-containing media through a 0.45 μm PES membrane syringe filter onto the MEF cells. Add polybrene reagent to virus-containing media on the MEF cells at a final concentration of 1 μg/mL. Replace media on Plat-E cells. Incubate MEFs with viral-containing media for 6–10 h. Repeat infection procedure and incubate MEFs with viral-containing media overnight.
8. Day 5: Repeat **step** 7, Subheading 3.1 beginning approximately 72 h after Plat-E cell transfection. Plat-E cells can be discarded after the fourth collection of viral-containing media (*see* **Note 5**).
9. Day 6: Approximately 96 h after Plat-E transfection, split the MEF cells and combine duplicate wells into a large dish to expand.
10. Test for expression: Plate some cells on a glass bottom dish or coverslip and image using a fluorescence microscope. For best

results in the fusion assay, approximately 95% of the cells need to be expressing the fluorescent protein (*see* **Note 6**).

11. Freeze several tubes at 2×10^6 cells per mL in freezing media. These can be thawed onto a 15 cm dish and split 2 days later.

3.2 Growing and Harvesting Cells

1. Begin cell growth for the assay: Three days in advance of the assay, seed five 15 cm plates with approximately 1×10^6 cells per plate (*see* **Notes 7** and **8**). To maintain these cells in culture, keep another plate growing (*see* **Note 9**). Populations needed will vary by experiment but minimally include two populations of wild-type cells, each expressing one of the two fluorescent proteins (five plates of matrix RFP and five plates of matrix CFP) (*see* **Note 10**).
2. On the day of the assay, make 25 mL MIB and store at 4 °C or on ice.
3. Working with the five plates for one cell population at a time, remove most of the media from dishes, leaving approximately 5 mL. Detach cells from the plates into the remaining media using a cell scraper (*see* **Note 11**).
4. Transfer the cells from all five plates for one population to one sterile 50 mL conical tube. Keep tubes on ice while collecting cells from the next set of dishes.
5. When all cells have been collected from all cell populations, centrifuge at $300 \times g$ for 10 min at 4 °C to pellet cells (*see* **Note 12**).

3.3 Isolation of Mitochondria

1. Return tubes to ice and aspirate media. Finger flick the tube to loosen the cell pellet and then resuspend cells in 5 mL cold MIB.
2. Pellet cells by centrifugation at $300 \times g$ for 5 min at 4 °C. Aspirate all buffer.
3. Resuspend cells in one pellet volume MIB and transfer to ice-cold homogenizers, one for each cell population (*see* **Note 13**). Cells will be homogenized twice sequentially and cell lysates from each homogenization combined prior to the mitochondrial isolation spin.
4. Homogenize cells on ice with a Teflon pestle operating at 400 rpm. Slowly move the pestle up and down 12–15 times, avoiding the formation of bubbles in the sample by keeping the pestle below the buffer at all times (*see* **Note 14**).
5. Transfer homogenate to a cold 1.5 mL microcentrifuge tube (Fig. 2, Tube A). Rinse the homogenizer with 150 μL MIB and combine with homogenate in microcentrifuge tube. Spin the homogenate at $300 \times g$ for 5 min at 4 °C to remove nuclei and unbroken cells.

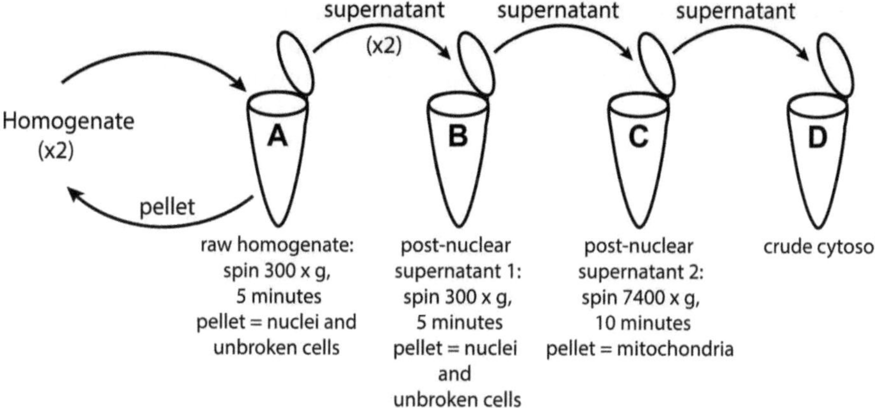

Fig. 2 Differential centrifugation overview beginning after homogenization

6. Transfer the supernatant to a new cold 1.5 mL microcentrifuge (Fig. 2, Tube B) (*see* **Note 15**).
7. Resuspend the cell pellet in one pellet volume cold MIB and put back into the same homogenizer from **step 4**, Subheading 3.2.
8. Homogenize on ice as in **step 4**, Subheading 3.2.
9. Transfer homogenate to the same 1.5 mL microcentrifuge tube from **step 5**, Subheading 3.2 (Fig. 2, Tube A).
10. Spin the homogenate at 300 × *g* for 5 min at 4 °C to remove any remaining unbroken cells and nuclei.
11. Combine this supernatant with the supernatant from **step 6**, Subheading 3.2 (Fig. 2, Tube B). Discard the pellet.
12. Spin the combined supernatant 300 × *g* for 5 min at 4 °C to remove any remaining unbroken cells and nuclei.
13. Transfer the postnuclear supernatant to a clean, cold 1.5 mL microcentrifuge tube (Fig. 2, Tube C). Discard the small pellet from **step 12**, Subheading 3.2. Spin the supernatant at 7400 × *g* for 10 min at 4 °C to pellet the mitochondria (*see* **Note 16**).
14. Transfer the post-mitochondrial supernatant to a cold 1.5 mL microcentrifuge tube and save as the crude cytosol fraction (Fig. 2, Tube D) (*see* **Note 17**).
15. Resuspend the mitochondrial pellet in 500 μL MIB and spin at 7400 × *g* for 10 min at 4 °C to wash the mitochondria (Fig. 2, Tube C) (*see* **Note 18**).
16. Aspirate all supernatant. Resuspend the mitochondrial pellet in one pellet volume MIB (*see* **Note 19**).
17. Determine the mitochondrial protein concentration by Bradford assay (*see* **Note 20**).

3.4 Fusion Assay

1. For each reaction condition, mix 10–15 μg of mitochondrial protein for each color of mitochondria in a 1.5 mL microcentrifuge tube. Make a master mix in MIB if the same mitochondrial pairing is being tested under multiple conditions and then aliquot mixed mitochondria into individual tubes (100 μL MIB per reaction in master mix).
2. Pellet the mixed mitochondria by centrifugation at 7400 × *g* for 10 min at 4 °C (*see* **Notes 21** and **22**).
3. Move the tubes to ice and incubate for 10 min.
4. Remove the supernatant from all tubes.
5. Add 10 μL of appropriate fusion reaction buffer to each tube on ice to cover the mitochondrial pellet. Fusion buffer as listed in Subheading 2 is the standard condition. Cytosol buffer can be used as a negative control. Crude cytosol (**step 14**, Subheading 3.3) or pure protein can be added to the standard fusion buffer to test the effect of these factors.
6. Resuspend mitochondria in fusion buffer by pipetting up and down, avoiding the formation of bubbles. Leave on ice until all tubes have been resuspended.
7. Incubate at 37 °C for 60 min (*see* **Note 23**).

3.5 Slide Preparation and Imaging

1. During fusion incubation, prepare slides for imaging.
2. Prepare agarose bed in single-well depression slides (Fig. 3a): Pipet 30–40 μL hot low melt agarose into the depression (Fig. 3b) and immediately put a plain glass slide on top to

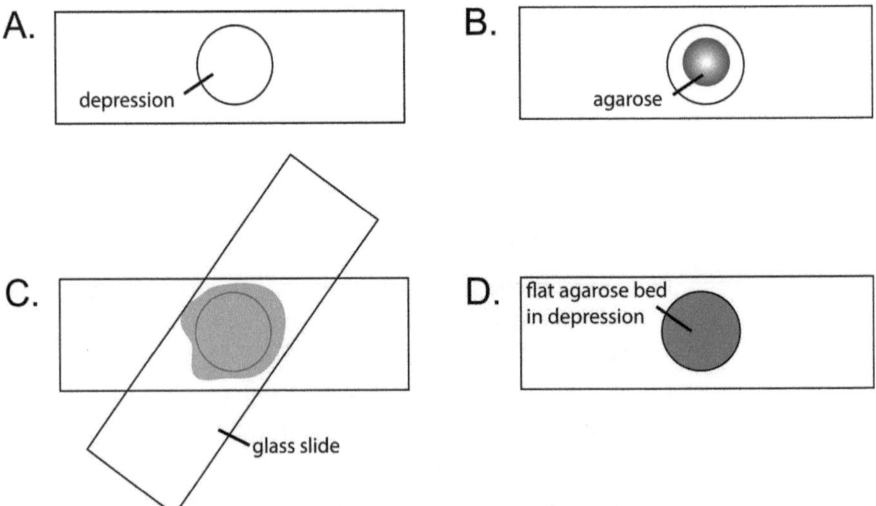

Fig. 3 Slide preparation schematic. (**a**) Illustration of single-well depression slide. (**b**) 3% low melt agarose in depression. (**c**) Plain glass slide placed over the hot agarose creating a flat agarose bed in the depression. (**d**) Final agarose bed in the depression after the top slide has been removed and extra agarose has been cleaned off

create a flat surface (Fig. 3c) (*see* **Note 24**). Just before use, move the top slide off of the agarose bed by pushing to the side (do not lift up). Clean off the excess agarose from around the depression with a razor and a delicate task wipe (Fig. 3d) (*see* **Note 25**).

3. At the end of the 60-min incubation, move tubes to ice.
4. Pipet 4 μL reaction mixture onto the agarose bed.
5. Drop a coverslip over the reaction mixture.
6. Seal the coverslip with nail polish.
7. Collect images of each reaction with a confocal or widefield fluorescence microscope with the appropriate excitation and emission settings to detect the matrix-targeted fluorophores. Refer to the guidelines below for further detail (*see* **Note 26**).
8. Collect Z stacks to capture the entire mitochondrial volume in both colors; we recommend 10–12 steps of 0.2 μm each (*see* **Note 27**). Collect multiple fields of view from different positions on the agarose bed so that at least 400 total mitochondria are imaged (Fig. 4a) (*see* **Note 28**).
9. Fusion efficiency is quantified by dividing the number of fused mitochondria (those with both fluorophores in three dimensions) by the total number of mitochondria in a given field of view (Fig. 4b). Count at least four fields and a minimum of 400 total mitochondria.
10. Analyze data by comparing the proportion of fused mitochondria among conditions. Due to daily variation in buffer components, temperature, etc., we normalize fusion efficiency to a control reaction of wild-type mitochondria with standard fusion buffer (*see* **Note 29**).

4 Notes

1. Use KOH pellets until the PIPES goes into solution, and then continue adjusting the pH with 6 M KOH.
2. Creatine kinase is a labile enzyme. Aliquot 1–3 mg of powder to microcentrifuge tubes upon receipt and note the weight of the powder on the tube. Store microcentrifuge tubes at 4 °C in a desiccator. On the day of the fusion assay, resuspend creatine kinase powder aliquot in cytosol buffer.
3. The pH of GTP and ATP can be measured using pH paper. It is important to adjust the pH of the GTP and ATP to 7.0 to maintain the stability of the triphosphate.

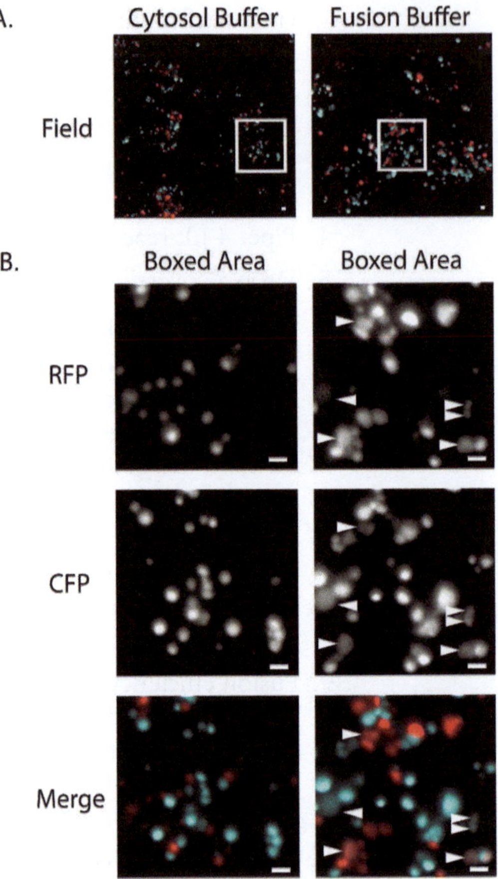

Fig. 4 Example of images of mitochondria after cell-free fusion assay performed in cytosol buffer as a negative control, or fusion buffer. (**a**) Field of mitochondria obtained at 100× by a widefield fluorescence microscope. (**b**) Detail of the boxed area in (**a**) showing mitochondria expressing RFP, CFP, and the merged image. Arrowheads indicate examples of fused mitochondria determined by overlap of RFP and CFP signal (scale bar = 1 μm)

4. We use three wells of a six-well dish for each color; i.e., wild-type RFP cells will require three wells of Plat-E cells and three wells of MEF cells.
5. By the fourth collection of viral supernatant, the Plat-E cells will look patchy and begin to lift off the dish.
6. Expression could also be tested and/or selected for by cell sorting.
7. Keep any necessary selection on the cells as they are being maintained and passed. The plasmids that we use for the mitochondrial marker fluorescent proteins do not encode a resistance gene for cell culture. When cells are expanded to five 15 cm dishes prior to fusion day, we do not maintain antibiotic selection.

8. Plating these dishes will require one confluent 15 cm dish of cells. The seeding density may need to be adjusted for the cell type to be utilized. On the day of mitochondrial isolation, cells need to be 90–100% confluent.
9. In addition to plates for the fusion assay, we maintain another set of plates for propagation and seeding fusion assay plates.
10. The wild-type populations are required every time as controls. Other cell types can be grown as necessary.
11. Cell scrapers can be washed in water and then in 70% ethanol and used again.
12. Spinning time can be adjusted depending on the appearance of the pellet. The pellet should be tight enough so that it does not come off the side of the tube. During this step, rinse homogenizer tubes in water and MIB and leave on ice.
13. About 200–400 μL.
14. The pestle is inserted into the overhead stirrer, and a glass tube is held in a small beaker of ice water while performing the strokes. It is imperative that the cells are concentrated at this point.
15. Supernatant should be cloudy.
16. Make cytosol buffer during this step.
17. This supernatant should be clear.
18. This removes any remaining cytosolic factors.
19. The mitochondrial pellet is usually 15–30 μL.
20. Create a standard curve and determine protein concentration according to the manufacturer's instructions. The mitochondrial concentration is usually 2–10 mg/mL.
21. Make working stocks of creatine kinase, creatine phosphatase, GTP and ATP, and then fusion buffer during this step.
22. This concentrates the mitochondria to establish proximity.
23. Other time points can be collected as well.
24. It might help to cut off the end of the pipet tip so the viscous agarose is easier to transfer. Make sure there are minimal bubbles in the agarose. Place the top slide before the agarose begins to solidify and at an angle so it can slide off easily (Fig. 3c).
25. The slides can be cleaned with 70% ethanol and reused indefinitely.
26. Isolated mitochondria vary in intensity of fluorescent signal, in part because fusion dilutes the fluorophores. When establishing your exposure conditions, ensure that the fluorescent signal is bright enough to visualize all mitochondria, especially those with lower intensity. This may result in some mitochondria with signal near saturation.

27. We use a 100× objective and collect fields of either 512 × 512 or 1920 × 1080 pixels.
28. Find fields with enough mitochondria that they are worth counting but not so crowded that you cannot distinguish one organelle from another (as in Fig. 4a).
29. Keep raw data along with the proportions that have been normalized to the wild-type fusion efficiency.

References

1. Smirnova E, Griparic L, Shurland D-L et al (2001) Dynamin-related protein Drp1 is required for mitochondrial division in mammalian cells. Mol Biol Cell 12:2245–2256
2. Santel A, Fuller MT (2001) Control of mitochondrial morphology by a human mitofusin. J Cell Sci 114:867–874
3. Olichon A, Baricault L, Gas N et al (2003) Loss of OPA1 perturbates the mitochondrial inner membrane structure and integrity, leading to cytochrome c release and apoptosis. J Biol Chem 278:7743–7746
4. Wakabayashi J, Zhang Z, Wakabayashi N et al (2009) The dynamin-related GTPase Drp1 is required for embryonic and brain development in mice. J Cell Biol 186:805–816
5. Ishihara N, Nomura M, Jofuku A et al (2009) Mitochondrial fission factor Drp1 is essential for embryonic development and synapse formation in mice. Nat Cell Biol 11:958–966
6. Chen H, Detmer SA, Ewald AJ et al (2003) Mitofusins Mfn1 and Mfn2 coordinately regulate mitochondrial fusion and are essential for embryonic development. J Cell Biol 160:189–200
7. Alavi MV, Bette S, Schimpf S et al (2007) A splice site mutation in the murine Opa1 gene features pathology of autosomal dominant optic atrophy. Brain 130:1029–1042
8. Davies VJ, Hollins AJ, Piechota MJ et al (2007) Opa1 deficiency in a mouse model of autosomal dominant optic atrophy impairs mitochondrial morphology, optic nerve structure and visual function. Hum Mol Genet 16:1307–1318
9. Waterham HR, Koster J, van Roermund CWT et al (2007) A lethal defect of mitochondrial and peroxisomal fission. N Engl J Med 356:1736–1741
10. Züchner S, Mersiyanova IV, Muglia M et al (2004) Mutations in the mitochondrial GTPase mitofusin 2 cause Charcot-Marie-Tooth neuropathy type 2A. Nat Genet 36:449–451
11. Alexander C, Votruba M, Pesch UEA et al (2000) OPA1, encoding a dynamin-related GTPase, is mutated in autosomal dominant optic atrophy linked to chromosome 3q28. Nat Genet 26:211–215
12. Delettre C, Lenaers G, Griffoin JM et al (2000) Nuclear gene OPA1, encoding a mitochondrial dynamin-related protein, is mutated in dominant optic atrophy. Nat Genet 26:207–210
13. Simula L, Campello S (2018) Monitoring the mitochondrial dynamics in mammalian cells. Methods Mol Biol 1782:267–285
14. Hoppins S, Edlich F, Cleland MM et al (2011) The soluble form of Bax regulates mitochondrial fusion via MFN2 homotypic complexes. Mol Cell 41:150–160
15. Mishra P, Carelli V, Manfredi G et al (2014) Proteolytic cleavage of Opa1 stimulates mitochondrial inner membrane fusion and couples fusion to oxidative phosphorylation. Cell Metab 19:630–641

Chapter 11

Electrophysiological Methods for Detection of Membrane Leakage and Hemifission by Dynamin 1

Pavel V. Bashkirov, Ksenia V. Chekashkina, Anna V. Shnyrova, and Vadim A. Frolov

Abstract

Membrane fusion and fission are indispensable parts of intracellular membrane recycling and transport. Electrophysiological techniques have been instrumental in discovering and studying fusion and fission pores, the key intermediates shared by both processes. In cells, electrical admittance measurements are used to assess in real time the dynamics of the pore conductance, reflecting the nanoscale transformations of the pore, simultaneously with membrane leakage. Here, we described how this technique is adapted to *in vitro* mechanistic analyses of membrane fission by dynamin 1 (Dyn1), the protein orchestrating membrane fission in endocytosis. We reconstitute the fission reaction using purified Dyn1 and biomimetic lipid membrane nanotubes of defined geometry. We provide a comprehensive protocol describing simultaneous measurements of the ionic conductance through the nanotube lumen and across the nanotube wall, enabling spatiotemporal correlation between the nanotube constriction by Dyn1, leading to fission and membrane leakage. We present examples of "leaky" and "tight" fission reactions, specify the resolution limits of our method, and discuss how our results support the hemi-fission conjecture.

Key words Lipid membrane nanotube, Patch-clamp, Fission, Dynamin 1, Membrane leakage

1 Introduction

Maintenance and functional coherence of endomembrane system depend on constant material exchange between isolated membrane compartments and organelles. The membrane recycling comprises two antagonistic processes: membrane fusion and fission [1]. Though these processes change membrane topology in opposite way, they both involve similar highly bent membrane structures [1–3]. Bending stress assists in local destabilization of lipid bilayer architecture leading to lipid rearrangements into the paradigm hourglass stalk structure, the intermediate shared by fusion and fission reactions [1]. Yet recent experiments and computer simulations showed that the stress could also cause formation of transient trans-membrane pores, thus revealing alternatives to the leakage-

Rajesh Ramachandran (ed.), *Dynamin Superfamily GTPases: Methods and Protocols*, Methods in Molecular Biology, vol. 2159, https://doi.org/10.1007/978-1-0716-0676-6_11,

free stalk mechanism of fusion and fission [4–6]. Though pore formation can compromise membrane barrier function, fundamental for cellular homeostasis, poration has long been linked to membrane remodeling [7–9], inferring that membrane leakage might be not a random product of high bending stress but rather a part of membrane remodeling mechanism, integrating the remodeling with other cellular processes, such as apoptosis [10].

Experimental assessment and analyses of leakage during fusion and fission require exceptional spatiotemporal resolution, as it first became evident upon application of an electrophysiological approach, patch-clamp admittance measurements, to exocytosis and viral fusion [11–15]. The measurements revealed that fusion begins form establishing a short-living narrow connection between the lumens of fusing compartments termed fusion pore [15–17]. Similar metastable intermediate was detected at the final stage of membrane fission [11, 18]. Real-time admittance measurements further showed that fusion and fission pores could open and close at sub-millisecond timescale, closely resembling ion channel behavior [11, 15]. Repetitive closures of the pore lumen, termed conductance flicker [11, 15], were later associated with a stress-driven structural instability leading to self-fusion of the inner monolayer of the pore [19], indicating that membrane leakage would transpire as similarly fast and stochastic process. Indeed, the flicker was found time-correlated with transient membrane leakage [20]. Nowadays, the admittance measurements still remain the technique of choice to reveal and study such time correlations.

The association of leakage with increased membrane bending stress in flickering pores is of special relevance for dynamins, a large superfamily of GTPase proteins widely implicated in intracellular membrane remodeling [21]. In-depth mechanistic analyses of several members of the family demonstrated that dynamin oligomers act by converting the energy of GTP hydrolysis into acute membrane curvature stress [22–24], potentially causing leakage. As each dynamin operates in a particular membrane (sub)system and tissue-specific isoforms of dynamins have been identified [21], dynamin-driven membrane leakage, or the lack of thereof, can have distinct physiological and mechanistic meaning. Specifically, multiple dynamins controlling fusion and fission of mitochondria membrane have long been linked to apoptotic protein machinery implicated in membrane poration and leakage [10, 25]. However, the mechanisms linking dynamic-driven membrane remodeling and leakage remain poorly explored.

Here we describe the methodology enabling a real-time correlative assessment of dynamin-driven membrane remodeling and leakage in vitro, in a simplified reconstitution system containing purified proteins and lipid membrane nanotubes [26, 27]. The methodology is based upon adaptation of the patch-clamp technique to lipid nanotubes, a generic biomimetic membrane template widely used for reconstitution of membrane remodeling activities of isolated proteins and protein complexes [28]. We found that glass microcapillaries used in the classical on-cell patch-clamp could be utilized to pull a lipid nanotube (NT) from planar lipid bilayer reservoir [11]. In this chapter, we explain how to produce the nanotubes and set up electrical measurements to simultaneously assess the ionic conductance through the NT lumen and across its wall. We further demonstrate how the technique was used to monitor in real time the NT constriction and membrane leakage by Dynamin1 (Dyn1), the founding member of dynamin superfamily orchestrating membrane fission in endocytosis [22, 24]. Dyn1 is a neuron-specific isoform so that plasma membrane leakage caused by the protein activity could lead to depolarization, potentially damaging for the cell function. We describe, step by step, how we reconstitute and quantify membrane remodeling, fission and leakage by Dyn1, address the sensitivity and major limitation of our method, and also discuss mechanistic interpretation of our findings, specifically hemi-fission mechanism of membrane remodeling by Dyn1. The protocol is not specific to Dyn1 and can be straightforwardly applied to assess membrane transformations and leakage produced by any protein or force-factor, e.g. osmotic pressure.

2 Materials

All solutions should be prepared with ultrapure water (18.2 MΩ·cm at 25 °C). The highest purity available reagents are to be used throughout the procedure. Protective equipment, safety instructions, and waste disposal regulations should be carefully followed, especially with organic solvents used in the procedure.

2.1 Planar Bilayer Lipid Membrane (BLM)

1. Lipids: 1,2-Dioleoyl-sn-glycero-3-phosphatidylcholine (DOPC), 1,2-dioleleoyl-sn-glycero-3-phosphatidylethanolamine (DOPE), 1,2-dioleoyl-sn-glycero-3-phosphatidyl-L-serine (DOPS); cholesterol (Chol), 1,2-dioleoyl-sn-glycero-3-phosphatidylinositol-4,5-bisphosphate ($PI(4,5)P_2$) from Avanti Polar Lipids (Alabaster, AL, USA).
2. Compressed gases: high-purity argon and nitrogen.
3. Solvents (all solvents should be analytical grade): octane, decane, squalane, chloroform, ethanol, methanol.
4. Teflon film, 25–50μm thickness.
5. Parafilm®.
6. Round Petri dishes, 35 mm in diameter.
7. BLM bathing solution (in mM): KCl:150, HEPES:10, EDTA:1, $MgCl_2$:2, pH 7.0. We routinely use KCl as

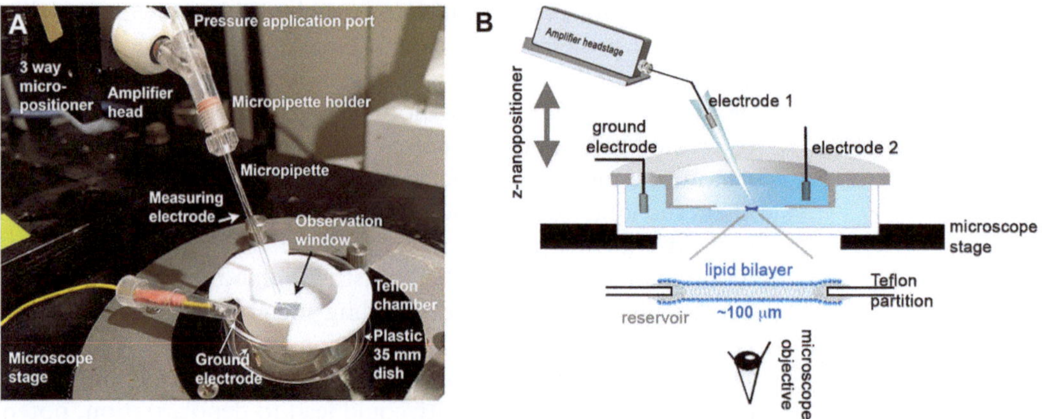

Fig. 1 Experimental setup for the NT production. (**a**) A photograph of the microscope stage showing the minimal set of components required for the NT production: the BLM chamber mounted on a Petri dish, the patch pipette in the micropipette holder, and the ground electrode placed into the Petri dish. (**b**) Schematic representation of (**a**) highlighting the 100 μm aperture with BLM and membrane reservoir

electrolyte, as similar mobility of K^+ and Cl^- in water facilitates interpretation of conductance measurements. Prepare monthly and store at 4 °C.

8. Amber glass vials with polytetrafluoroethylene (PTFE) caps.
9. Kolinsky sable brushes 3/0.
10. Vacuum pump.
11. BLM chamber (*see* **Note 1**). We use a home-built circular Teflon® chamber that can be mounted on top of a standard 35 mm Petri dish (Fig. 1a, [29]). The bottom of the chamber has ~5 × 5 mm observation window covered by a thin (25–50 μm) transparent Teflon® film (called partition). The partition is attached by PDMS, molten Parafilm®, or an adhesive and is to be replaced bimonthly [29]. The partition has a small (50–100 μm) circular aperture for the BLM formation (BLM aperture) produced by a sharp microneedle or a puncher (Fig. 1a, b).
12. Brush/BLM chamber cleaning station: 50 mL glass vials containing methanol and distilled water, empty vials for the waste solutions.

2.2 Micropipette Fabrication and Filling

1. Borosilicate glass or quartz capillaries for micropipettes preparation (World Precision Instruments, USA or Science Products, Germany).
2. Micropipette puller (Model P-2000, Sutter Instruments Co., USA).
3. Basic inverted microscope to monitor the pipette tip.
4. MicroFil® needles, type MF34G.

5. 1 mL plastic syringes.
6. 5 μm plain silica beads (from Microspheres-Nanospheres, USA).
7. Pipette filling solution, same as BLM (in mM): KCl:150, HEPES:10, EDTA:1, $MgCl_2$:2, pH 7.0. Prepare monthly and store at 4 °C.

2.3 Patch-Clamp Conductance Measurements

1. Anti-vibration table (KSI 9100, Kinetic Systems Inc., USA) or similar.
2. Inverted microscope (e.g., Eclipse Ti-U, Nikon, Japan) with 40× long working distance objective. Using phase-contrast optics is desirable, addition of high-power objective length (60× or 100× oil) and epi-fluorescence will enable several essential control procedures (nanopositioner calibrations, control of the protein delivery).
3. Imaging module (video-camera and a monitor, e.g., Nikon DS system, Nikon, Japan).
4. Two patch-clamp amplifiers (e.g., Axopatch 200B, Molecular Devices, LLC, USA).
5. Patch-clamp micropipette holders with a measuring electrode (e.g., from *ALA* Scientific Instruments, USA, most patch-clamp amplifiers are sold with a suitable holder).
6. Patch-clamp micropipette holder without the electrode for the protein delivery pipette. Note that the pressure port of this holder is to be sealed.
7. Two high-quality 3D micropositioning system (referred to as a "micromanipulator" below) consisting of three-axis linear stage (with at least 1 cm displacement range in each axis) equipped with three drivers (Sensapex Oy, Finland, or similar). The stages and the drivers should have minimal temperature and mechanical drifts. One of the stages should be switchable between coarse and extremely fine displacement modes in vertical direction. For that precise nanopositioner (Newport ESA-CSA, Newport Corp., USA, resolution >20 nm) can be mounted sequentially with the coarse driver.
8. Suitable connectors for firmly attaching patch-clamp amplifier head-stage and the delivery pipette holder to the three-axis stage.
9. Ag/AgCl pellet electrodes for ground and second measuring electrodes (Fig. 1b), and silver chloride bath electrode prepared from Teflon-coated silver wire for first measuring electrodes (World Precision Instruments, USA).
10. Adjustable magnetic holders (Bioscience Tools, USA).
11. Digital oscilloscope (e.g., Tektronix, TDS-2024B, USA).

12. Signal generator (GOS-620FG, Good Will Instrument Co., Ltd., China, or similar).
13. Data acquisition system: ADC converter (Digit Data 1550, Molecular Devices, LLC, USA) and appropriate software packages (pClamp 9.2, Molecular Devices, LLC, USA).
14. Software package for off-line data analysis (e.g., OriginLab, https://www.originlab.com).

2.4 Proteins and Nucleotides

1. Purified dynamin 1, frozen in small aliquots (*see* **Notes 2** and **3**).
2. High purity GTP, preferably Li salt.
3. Small volumes (10–20 μL) dialysis device.

3 Methods

3.1 The Patch-Clamp Setup

3.1.1 Setup Assembly

The microscope is to be set on a high-quality vibration isolation station to minimize jittering of the patch-pipette tip. The micromanipulators are mounted onto or near the microscope stage. The head-stage of the main patch-clamp amplifier (Fig. 1a) is attached to the 3D micropositioner with a fine-control vertical displacement. The holder for the delivery pipette is mounted on the second micropositioner so that the delivery pipette can reach the center of the partition aperture. Ground electrode is attached to the microscope stage using adjustable tilt magnetic holder for easy insertion into the Petri dish (Fig. 1b). If the second patch-clamp amplifier is used, its head-stage with the second measuring electrode shall be mounted near the BLM chamber (on the microscope stage or a separate platform located above the stage) in such a way that the second measuring electrode can be introduced into the upper compartment of the BLM chamber (the protocol parts requiring the second amplifier are indicated below).

3.1.2 Electrical Noise Minimization

All of the metal parts of the setup, especially the microscope body and micromanipulators, should be properly grounded [30]. The area near the experimental chamber should be shielded from the external electrical noise by a Faraday cage. Instead of a large cage enclosing the whole setup, we prefer local shielding constructed around the microscope stage. With open input of the amplifier, the current output should be a flat baseline with the background noise amplitude of ~0.1 pA at 100 mV/pA amplifier gain.

3.1.3 Definition of the Main Experimental Observables

The outputs of the patch-clamp amplifier and nanopositioner controller shall be connected to the data acquisition system to record the following observables:

1. Electrical potentials applied to the measuring electrodes.
2. Electrical current at the measuring electrodes.

3. Z-coordinate of the patch pipette tip (from readings of the vertical nanopositioner and/or calibration curve, *see* **Note 4**).

Chose adequate sampling frequency [22, 24]. 1 kHz is generally sufficient, but it is advisable to increase the sampling rate tenfold when the task is to observe rapid events such as membrane poration (leakage) or fission pore flickering [27, 31].

3.2 Micropipette Preparation

3.2.1 Micropipette Fabrication

Thin-wall borosilicate glass is generally the best choice for the BLM experiments (*see* **Note 5**). Refer to a micropipette puller instruction (e.g., Sutter Pipette Cookbook for Sutter micropipette pullers) for the corresponding micropipette fabrication protocol enabling reproducible produciton of micropipettes with the tip diameter of ~1 μm from this glass capillaries (*see* below). Ideally, the micropipettes should be fabricated just before the experiment. If the pipettes are produced in advance, they should be kept in a sealed jar to minimize contamination on the pipette tip.

3.2.2 Micropipette Filling

The micropipette is backfilled with a desired electrolyte solution (BLM bathing solution) using a MicroFil® needle attached to a plastic syringe (*see* **Note 6**). Check for small air bubbles in the pipette tip using an inverted microscope. If found, try refilling or shaking the bubbles out by gentle tapping on the upper part of the micropipette.

3.3 Planar BLM Formation

3.3.1 Preparation of Lipid Solutions

Two solutions of the same lipid mixture should be prepared, one for the pre-treatment of the BLM aperture and another for the "painting" of the BLM. Start by mixing lipid stocks in chloroform in two clean amber-glass vials at a desired molar ratio (e.g., use DOPC:DOPE:Chol:DOPS:PI(4,5)P_2 at 29:20:30:20:1 mol% as the reference lipid membrane composition to study Dyn1-mediated membrane fission [26]). Evaporate chloroform using gentle nitrogen gas stream followed by vacuum desiccations for 30 min to 1 h. To make the "pre-painting" solution, redissolve the lipids in one of the vials in 1:1 decane:octane (vol/vol) mixture to reach the final lipid concentration of 10 g/L. For the "painting" solution, add squalane to the second vial to reach the final lipid concentration of 10–30 g/L. Prepare 30–60 μL of each solution. Fresh preparations are preferable, although the mixtures can be stored at −20 °C for a week. To prevent lipid oxidation, the vials containing the pre-painting and the BLM solutions should be briefly purged with argon gas and tightly closed for storage.

3.3.2 Making the BLM

Note that the multistep procedure presented here is specific for the "painting" method (*see* **Note 1**).

Pre-treatment of the BLM Aperture

The first step in the BLM preparation is the deposition of a small amount of the lipid mixture of interest on the rim of the aperture in the Teflon film. A small (1–2 μL) drop of the "pre-painting" solution is gently placed on the aperture to cover it completely, then dried under a N_2 stream until the circular layer of the dried lipid film is observed around the aperture.

Assembly of the BLM Chamber

Upon the pre-treatmen of the partitiont, the teflon chamber (Fig. 1a) is settled on top of a 35 mm Petri dish, so that the partition separates the two compartments, the upper and the lower (Fig. 1b).

Addition of the Bathing Solution

The upper and lower compartments should be filled to the same level to avoid hydrostatic pressure difference across the partition aperture.

Electrodes Setup (*See* **Note 7**)

Insert the ground electrode into the lower compartment. When needed, insert the second measuring electrode into the upper compartment, apply zero holding potential to it, and then compensate the second and the ground electrodes offset. Next, immerse the micropipette containing the first measuring electrode into the upper compartment of the BLM chamber. Use low (1 mV/pA or lower) gain mode of the patch-clamp amplifier to avoid overloading the amplifier. Apply triangle wave potential (e.g., from −10 to +10 mV, ramp of 1 V/s) to the principal measuring electrode.

Measuring of the Patch-Pipette Resistance

Note that at this stage the chamber, and hence the micropipette interior, is electrically connected to the ground electrode in the Petri dish through the aperture in the Teflon film (over which the BLM will be painted). If the pipette is properly connected to the BLM chamber (that is, the pipette tip is not clogged), the amplifier should report a triangle wave current. Apply zero holding potential and compensate the first and the ground electrodes offset [29, 30]. Next, apply a small holding potential ([30], $V_h \sim 10$ mV) and measure the micropipette resistance, which at $V_h = 10$ mV and physiological ion strength of the bathing solution should be in the range of 3–6 MΩ. Then again apply saw-tooth wave potential (V_{st}, e.g., from −10 to +10 mV, with the ramp rate ~1 V/s).

"Painting" of the BLM

The procedure has been extensively described elsewhere (e.g., *see* ref. 29). Briefly, collect a small amount of the "painting" lipid solution (3.3.1) by the tip of a clean Kolinsky sable brush. Introduce the brush into the upper compartment of the chamber and, with the help of (video) microscopy, gently smear the solution over the BLM aperture to form a thick film covering the aperture. In a two-electrodes scheme (when the second electrode is out of the upper compartment of the chamber), formation of the thick film is seen as transformation of the triangle wave signal (conduction current) into the characteristic pulse-wave capacitive current. Next, the thick lipid film should thin spontaneously pushing the excess material towards the aperture rim where it forms a visible thick Plateau-Gibbs border ("reservoir," Fig. 1b). Of note, this membrane reservoir maintains the lateral tension of the lipid bilayer constant throughout the experiment. Inside the border, a thin and transparent lipid bilayer (the BLM) forms. During the bilayer formation, the amplitude of the pulse-wave current signal measured by the first measuring electrode should gradually increase. If the thick film remains stable and no current amplitude increase is observed, remove some of the deposited lipid mixture using a clean brush.

3.3.3 Checking of the BLM Quality

It is essential to confirm that the film thinning yields a true lipid bilayer in the aperture. First, determine the area S of the BLM within the reservoir boundary. Second, measure the total electrical capacitance (C) of the BLM and teflon partition separating the upper and lower compartments of the BLM chamber (Fig. 1b): C is proportional to the amplitude of the capacitive current (I_{cap}) as:

$$C = \frac{I_{cap}}{dV/dt},$$

where dV/dt is the voltage ramp rate (Fig. 2b). As lipid bilayer is much thinner than the partition, C effectively reports the total capacitance of the lipid film. Calculate the specific capacitance of the BLM film $C_{sp} = C/S$. For squalane-based BLMs, C_{sp} shall be close to 1 $\mu F/cm^2$ (e.g., 0.93 ± 0.17 $\mu F/cm^2$ for the lipid composition used here, *see* also **Note 8**). Smaller values indicate that the film remains thicker than lipid bilayer. It is also strongly advised to check the stability of the BLM under elevated holding potentials (100–200 mV) used in the NT experiments.

3.4 Formation of Lipid Nanotubes

3.4.1 Patching the BLM

Using the coarse micromanipulator, lower the micropipette so that its tip approaches the BLM center (Fig. 1b). Visualize the pipette approach via phase-contrast or bright-field microscopy (Fig. 2a). Stop the tip 10–20 μm above the BLM (Fig. 2a) and use the nanopositioner for the final slow approach. Stop as soon as the pipette touches the BLM and mark the reading of the nanopositioner as the BLM position (*see* **Note 4**). The moment the pipette touches the BLM can be determined either visually using phase-contrast microscopy or through the characteristic abrupt reduction of the capacitive current, which then gradually decreases to zero forming the tight contact between the BLM and the pipette tip (Fig. 2b).

3.4.2 Breaking the Patch

Once the tight contact (termed giga-seal) is formed, the membrane patch, now isolated inside the micropipette, should be ruptured in such a way that the pipette tip remains in contact with the lipid bilayer [11, 29]. This can be achieved by applying negative pressure

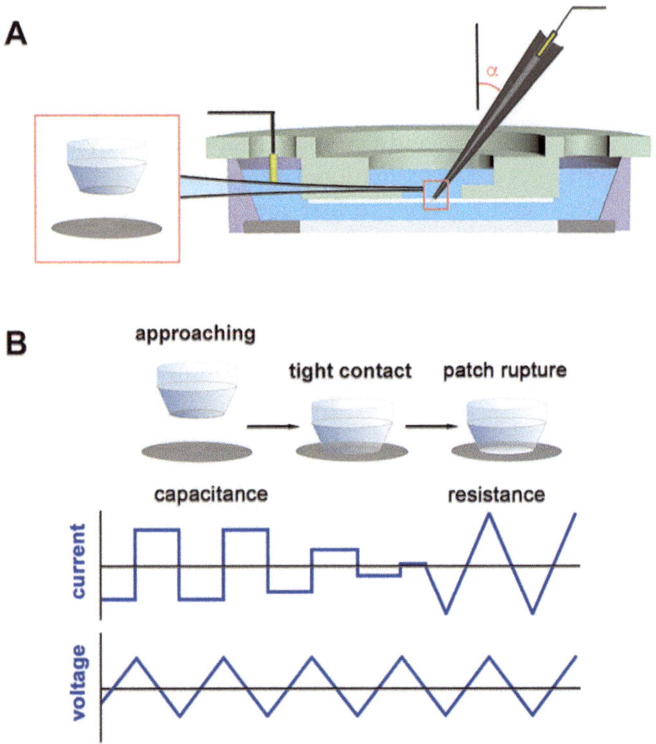

Fig. 2 Patch-clamping of the BLM. (**a**) Cartoon illustrating the position of the patch-pipette before the final slow approach to the BLM. (**b**) Cartoon illustrating changes in the electrical signal during the approach and formation of the tight contact between the patch-pipette and the BLM followed by the patch rupture

or a high voltage transient (using ZAP function of the patch.clamp amplifier) to the patch pipette interior. Once the membrane patch is ruptured, the triangle wave conduction current should reemerge (Fig. 2b).

3.4.3 Making the NT

Switch off the triangle wave potential and apply 10–50 mV constant V_h to the first measuring electrode. Slowly (0.1–0.5 μm/s) move the pipette up (away from the BLM) using the nanopositioner. Record the nanopositioner readings as well as changes in the current through the first measuring electrode (hereafter I_m). At the beginning of the pipette displacement, I_m is determined by the patch pipette conductance. With the pipette moving up, hourglass-shaped (catenoid) membrane bridge forms between the pipette tip and the BLM and I_m declines gradually as the catenoid thins with elongation [11] (Fig. 3a). Upon approaching a critical length (L_c), the catenoidal neck destabilizes and quickly shrinks ("collapses," [11, 29]). The collapse is seen as an abrupt drop in I_m (Fig. 3a). Upon detecting the collapse, stop the pipette movement.

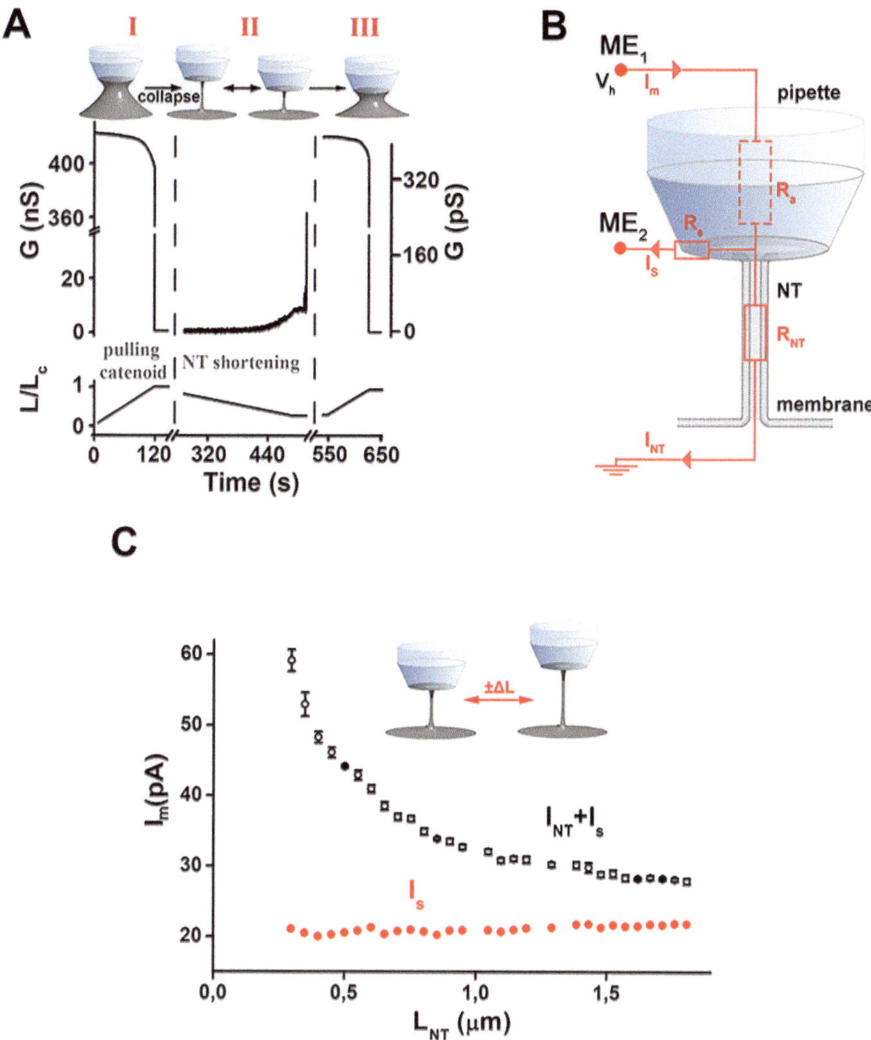

Fig. 3 Production and initial characterization of the NT. (**a**) Cyclic transformations between catenoidal neck and the NT (cartoon) caused by patch-pipette movement to/from the BLM (reproduced from ref. 11, copyright (2003) National Academy of Sciences). (**b**) An equivalent electrical circuit used in the NT conductance measurements; ME 1 and 2 are the measuring electrodes. (**c**) Changes of I_m and I_s during NT elongation

3.4.4 Verifying the NT Formation

The conductance drop (Fig. 3a) marks the transformation of the catenoidal tube into a thin cylindrical NT having several orders of magnitude less electric conductance than the catenoidal tube (Fig. 3a). However, accompanying stresses produce NT breakage in 10–30% of the cases, dependently on the lipid composition (*see* **Note 9**). In principle, membrane nanotubes can be detected by phase-contrast or fluorescence microscopy [27, 32]. However, for vertically oriented NTs used in this protocol such detection is problematic. Then, the NT formation is to be verified by the same conductance measurements. Albeit very narrow, the NT lumen (whose diameter varies from few to tens of nm) closely resembles a cylindrical nanopore filled with an electrolyte solution. The conductance properties of such nanopores are well understood [33]. The main concept of this protocol is to use the luminal conductance, G_{NT}, as the reporter of the NT appearance and the measure for changes of the NT geometry and topology. Following Ohms law for cylindrical conductor:

$$G_{NT} = \frac{\pi r_{NT}^2}{\rho_{NT} L_{NT}}, \tag{1}$$

where ρ_{NT} is the specific electrical resistance of the electrolyte filling the NT lumen, the resistance depending on the bulk electrolyte concentration and the membrane surface charge (*see* **Note 10**), and r_{NT} and L_{NT} are the radius of the NT lumen and the NT length correspondingly. I_m, proportional to G_{NT}, shall increase hyperbolically when the pipette moves towards the BLM (Fig. 3a). The absence of the I_m increase indicates that the NT failed to form (*see* **Note 9**). In such a case, gently (to avoid BLM rupture) move the patch-pipette up and from the solution, remove the pipette from the holder and discard. Repeat steps in Subheadings 3.2, 3.3.2, 3.3.3 (note that to measure the pipette resistance the second measuring electrode (Fig. 1b) shall be installed) and then repeat Subheading 3.4. If I_m starts increasing, keep moving the pipette towards the BLM, go extremely slow (0.1 μm/s) once the pipette is near (<1 μm) the BLM (its position determined in Subheading 3.4.1). At some point, the NT should spontaneously "open up," transforming back to the catenoidal neck (Fig. 3a). It is advisable to perform at least one cycle of such catenoid-to-NT transformations to ensure that the contact between the patch-pipette rim and the membrane is stable during the NT manipulations.

3.4.5 Simultaneous Detection of G_{NT} and Membrane Leakage

The holding potential V_h is applied to the first measuring electrode inside the patch pipette, with the pipette interior acting as an access resistance (R_a [30], Fig. 3b) to the upper end of the NT. The lower end of the NT facing the Petri dish is grounded. As R_a is much smaller than the integral resistance of the NT lumen (R_{NT}) (MΩ vs. GΩ), V_h defines the electrical potential difference between the NT ends. The current through the first measuring electrode (I_m) sums up the NT transluminal current and the leakage current between the patch-pipette interior and the BLM chamber, going predominantly through the giga-seal contact resistance (R_s, Fig. 3b). This leakage current through the giga-seal contact (I_s) is measured by the second measuring electrode. If this electrode is held at zero potential, G_{NT} can be straightforwardly determined as $G_{NT} = (I_m - I_s)/V_h$. Should membrane poration appear, it will be seen as an increase of I_s. However, with the pore location undetermined, I_s increase only reports the leakage with the sensitivity of the detection depending on the nanotube length (*see* Subheading

3.5.6). Notably, without poration I_m but not I_s changes with the patch-pipette movement (Fig. 3c), demonstrating that the giga-seal resistance remains stable during changes of the NT length.

3.4.6 Measuring G_{NT} with a Single Measuring Electrode

When the NT length is increased above 10 μm, its luminal resistance exceeds that of the giga-seal so that $I_m(\infty)$ approaches I_s (Fig. 3c). Hence, with just the first measuring electrode in action, I_s can still be measured at infinite NT extension (in practice, at $L_{NT} > 20$ μm at $V_h = 50$ mV) so that $G_{NT} = (I_m(\infty) - I_s)/V_h$.

3.5 Quantifying the NT Constriction and Fission by Purified Dynamin 1 or Osmotic Pressure

3.5.1 Protein Preparation

Thaw a Dyn1 aliquot, dialyze against the BLM bathing solution, keep on ice, discard at the end of the day of the experiment.

3.5.2 Making Delivery Pipette

The delivery pipette is produced the same way as patch-pipettes used for NT pulling. The tip of the delivery pipette shall be slightly wider, with the corresponding electrical resistance of 0.5–1 MΩ. Fill the pipette with the Dyn1 solution right before using it. Fill the tip first, then backfill to the level defined by the capillary force (*see* **Note 6**). Extra care should be taken to exactly match this level to avoid water fluxes in/from the pipette that interfere with the NT constriction measurements. Ideally, Dyn1 shall leave the pipette by passive diffusion only.

3.5.3 Protein Delivery to the NT

Upon formation and quantification of the NT geometry, Dyn1 could be added directly to the upper compartment of the BLM chamber. Changes of the NT conductance are monitored in the high-gain mode (100 mV/pA) of the patch-clamp amplifier, $V_h = 100$ mV. However, the protein addition to the bulk requires relatively large amounts of Dyn1 and might also compromise the stability of the BLM. Hence, localized perfusion of the NT by Dyn1 -ontaining solution is recomended. To supply Dyn1 to the NT visinity, carefully bring the delivery pipette close to the patch-pipette holding the NT (Fig. 4a). The local addition of Dyn1 produces a characteristic gradual decrease of the nanotube conductance till a new stationary level (Fig. 4b). The degree of narrowing of the lumen of NT could be expressed in dimensionless quantity (G_n)—conductance normalized on its initial value. Further elongation of the NT triggers a new constriction phase (Fig. 4c), confirming that Dyn1 is constantly delivered to the NT from the delivery pipette.

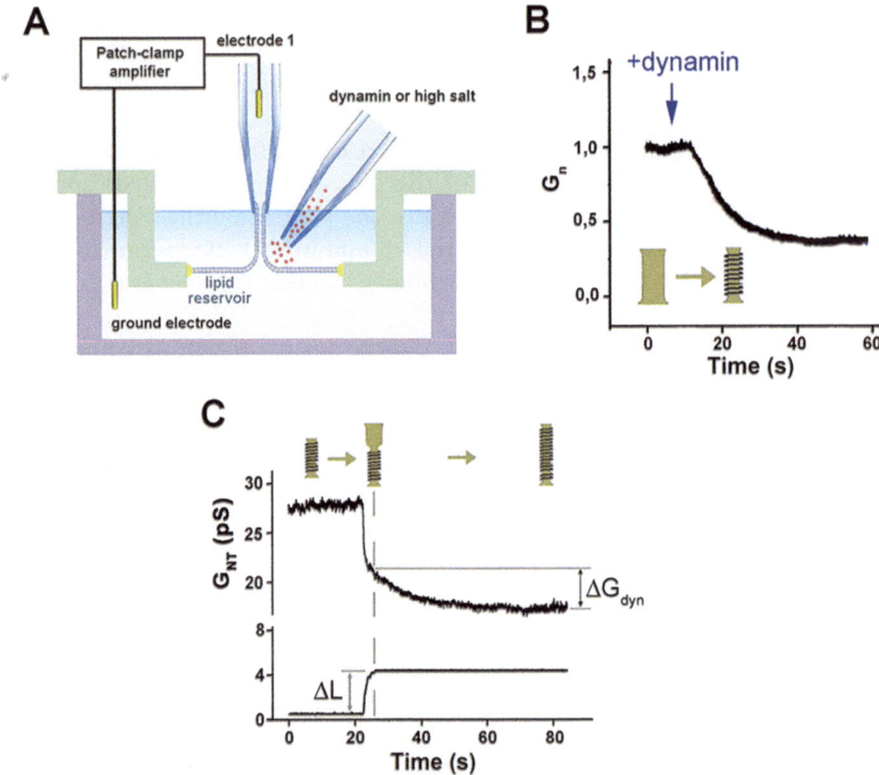

Fig. 4 Monitoring of the NT constriction. (**a**) Cartoon illustrating Dyn1 delivery at the vicinity of the NT using a delivery pipette. (**b**, **c**) Changes of the NT conductance during Dyn1-driven constriction of a static NT (**b**) and upon the NT elongation (**c**). (Reproduced with permission from ref. 26)

3.5.4 Quantifying the Uniform NT Constriction

Dyn1 (as many other curvature-creating proteins) imposes stable cylindrical membrane geometry in the absence of GTP. The Dyn1-driven constriction of a cylindrical NT is measured as the change of the cylinder radius (Fig. 5a), following Eq. 1. Crucially, the final radius of the constricted tube does not depend on the NT length (Fig. 5b), confirming uniform cylindrical constriction by Dyn1. The radius r_{Dyn} (Fig. 5b) thus characterizes the intrinsic curvature of the Dyn1 helix. Note, however, that r_{Dyn} is the radius of the NT lumen; the lipid bilayer thickness (4 nm) is to be added to r_{Dyn} to obtain the radius of the inner surface of the Dyn1 helix. As r_{Dyn} closely approaches the Debye length (λ), the impact of the electrical double layer should be taken into account (*see* **Note 10**).

3.5.5 Detecting Hemi-Fission

Without nucleotide, Dyn1 does not produce NT fission. To trigger the fission reaction, GTP (1 mM) has to be added both to the BLM compartment and to the protein delivery pipette. The simultaneous addition of Dyn1 (0.5 μM of Dyn1 in the delivery pipette) and GTP produces an abrupt drop in G_{NT} on average 30s after the protein addition (Fig. 6a, [26, 27]). Following the drop, G_{NT} stabilizes near background/zero level and does not increase when

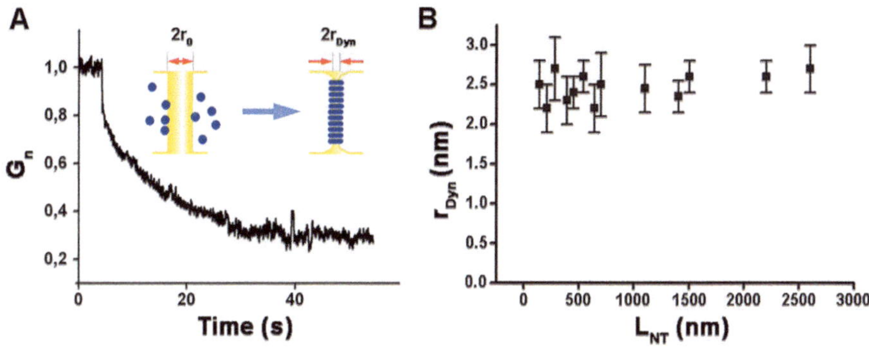

Fig. 5 Quantifying the NT curvature changes made by Dyn1. (**a**) The NT conductance decreases to a new stationary level corresponding to the formation of the protein scaffold uniformly covering the NT. (**b**) The NT radius measured in the constricted state (r_{NT}, *see* **a**) does not depend on the NT length measured before the Dyn1 addition. (From ref. 27. Reprinted with permission from AAAS)

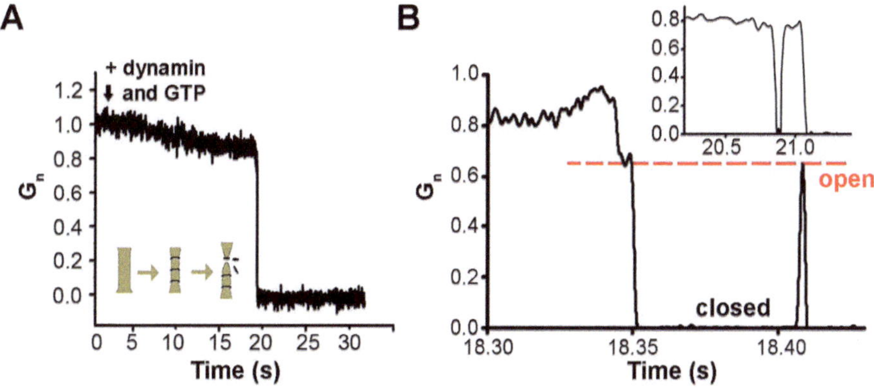

Fig. 6 Detection of the hemi-fission membrane transformation by conductance measurements. (**a**) Changes of the NT conductance upon addition of Dyn1 in the presence of 1 mM GTP (reproduced with permission from ref. 26). (**b**) Examples of reversible hemi-fission transformation where the NT stays in the hemi-fission (closed) state for a long time before briefly returning to an open configuration corresponding to the constricted NT. The insert shows a short-living closed state. (From ref. 27. Reprinted with permission from AAAS)

the patch-pipette is moved closer to the BLM. This loss of the luminal conductivity indicates hemi-fission [19, 26, 27]. The hemi-fission transformation can be reversible, seen as a transient resurgence of measurable G_{NT} (Fig. 6b). Note that the detection of fast reversible jumps of G_{NT} (termed "flicker") might require high-speed (10 kHz sampling) acquisition mode of the patch-clamp amplifiers.

3.5.6 Detection of Membrane Poration During NT Constriction

The bending stress increase due to the NT constriction could destabilize the NT membrane. We demonstrated earlier that the NT constriction driven by osmotic pressure led to upward jumps of I_m resembling protein channel openings ([31], Fig. 7a). Such jumps indicate membrane poration. To confirm that the pores are formed in the NT membrane, a second electrode should be

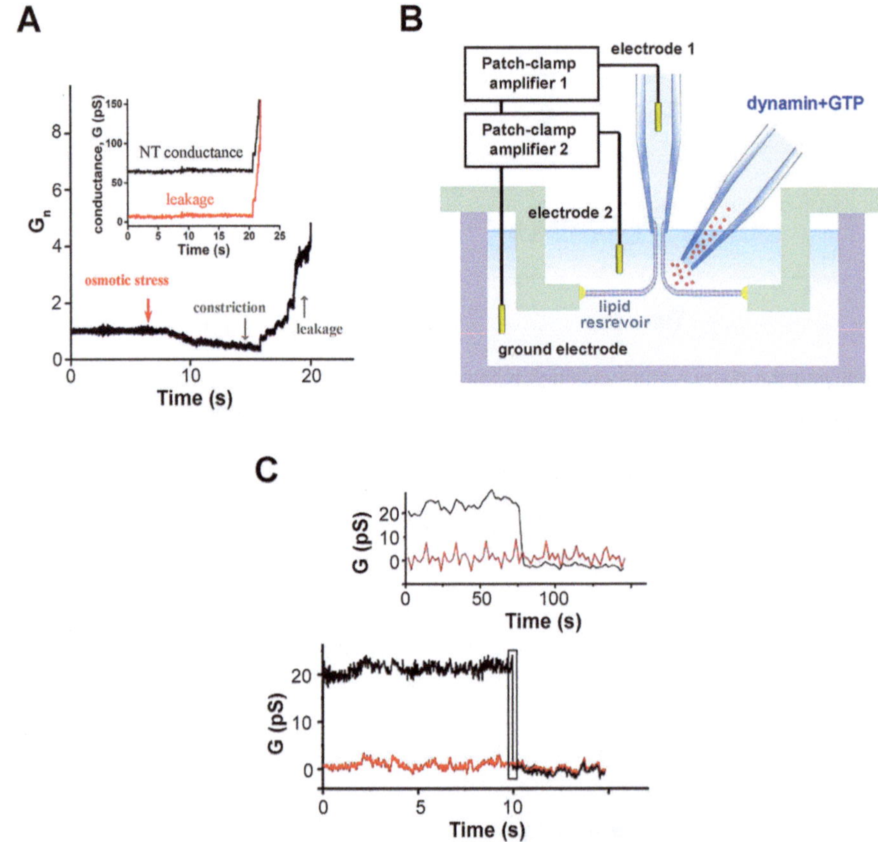

Fig. 7 Detecting membrane poration during NT constriction and fission. (**a**) The upward changes of I_m, indicating membrane poration, detected upon the NT constriction by osmotic stress (red arrow) [31]. The inset shows that the pore formation (membrane leakage) leads to synchronous conductance increase measured by the first and the second measuring electrodes. (**b**) Cartoon illustrating the usage of two measuring electrodes for detection of membrane poration. (**c**) Lack of changes in I_s reporting poration absence during Dyn1-driven hemi-fission. (Reproduced with permission from ref. 26)

introduced to the upper compartment of the BLM chamber (Fig. 1b) to measure the conductance between the NT interior and the external solution (Fig. 7b). Simultaneous recording of I_s and I_m demonstrate synchronous current increase indicative of the leakage in the NT membrane (Fig. 7c, insert). Crucially, no changes in I_m were detected during Dyn1-driven fission (Fig. 7c), corroborating leakage-free NT transformation along the hemi-fission pathway [19, 26, 27] (*see* **Note 11**).

4 Notes

1. Other methods for horizontal planar lipid bilayer preparation can be used. Note that the BLM should be easily accessible from above to perform patch-clamp manipulaitons.

2. Conventional procedures for protein purification in bacterial or insect cell can be used. Here, Sf9 cells were transfected with DNA encoding wild-type human dynamin 1. Proteins were purified by affinity chromatography using glutathione S-transferase (GST)-tagged Amphiphysin-II SH3 domain as the affinity ligand. Purified proteins were dialyzed overnight in 20 mM HEPES (pH 7.5), 150 mM KCl, 1 mM EDTA, 1 mM DTT, and 10% (v:v) glycerol, aliquoted (a single 20 μL × 10 μM aliquot is sufficient for a day of NT experiments), flash-frozen in liquid N_2, and stored at −80 °C. Protein concentrations were determined by absorbance at 280 nm using 56,185 M^{-1} cm^{-1} molar extinction coefficient.
3. As NT experiments are quite involved, it is strongly advisable to perform basic functional tests for each newly purified protein batch, such as giant vesicle tubulation and GTPase activity tests. Also, it is desirable to check whether Dyn1 from a particular protein batch does not interfere with the BLM stability. After completing step in Subheading 3.3.3 add the protein (0.5–1 μM) to the BLM chamber (upper compartment). Apply 100 mV holding potential to the first measuring electrode and record I_m changes. If detecting instabilities caused by the protein addition, such as baseline drift, pore-like events, or BLM rupture, discard the protein batch.
4. Precisely controlled vertical displacement of the pipette is critical for the method. Make the calibration curve for the nanopositioner to verify that the pipette tip properly follows commands of the nanopositioner controller (Fig. 8). To make the calibration, attach a high-contrast object (a point scatterer) to the pipette tip. We attach a 200 nm fluorescent microsphere to the patch-pipette tip and measure its displacement in a vertical direction using fluorescence microscopy with high-resolution optics (100× 1.4NA TIRF objective). The same optical setup can also assess the long-term positional stability of the pipette.
5. BLMs containing high amounts of charged lipid species and/or cholesterol can be difficult to patch using conventional glass capillary design for on-cell experiments. Using thick-wall glass capillary or quartz capillary can help to overcome the problem.
6. Fill the pipette to the level defined by the capillary raise to avoid water fluxes through the tip. To determine the level, dip the capillary tube (used for the patch-pipette preparation) into the bathing solution and measure the height of the capillary rise.
7. Set up the electrode and measure the pipette resistance as quickly as possible as the pre-painting slowly goes off the BLM aperture upon hydration.

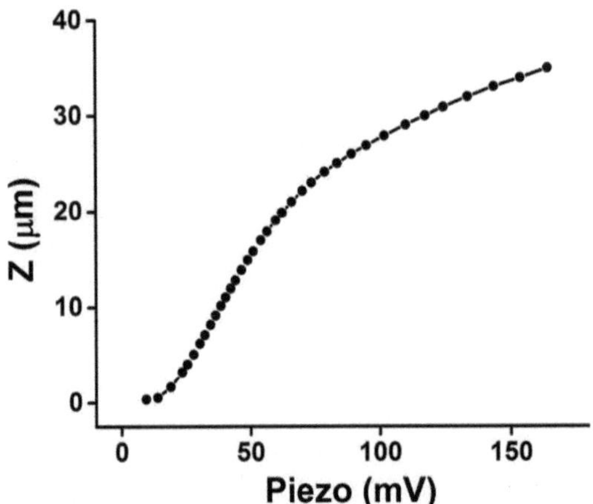

Fig. 8 An example of the calibration curve linking the readout from the nanopositioner controller (piezo) with the vertical displacement of the patch-pipette

8. The specific capacitance of squalane-based "painted" BLM is similar to that of the "solvent-free" BLMs produced by the Montal technique.
9. The major source of the NT breakage during the collapse is coarse pipette movements. Check the intrinsic vibration of the pipette using high-power (60 or 100× objective) phase-contrast microscopy and improve the anti-vibration strategy if large jittering is seen. Using the same optics check whether the nanopositioner causes the pipette vibrations. Also, note that the angle α between the pipette axis and the membrane normal shall be between 70° and 90° (Fig. 2a). With larger deviation from the vertical position of the patch pipette, the NT is likely to form near the rim of the pipette tip; the NT end can migrate into a "glass" area, thus blocking electrical access to the NT lumen. No detectable changes in I_m during the pipette movement might also indicate a low resolution of your I_m measurements. Improve the signal/noise ratio. Increase the amplifier gain. While low (1 mv/pA) gains are to be used up to the catenoidal neck collapse, NT parameters are to be measured with high (>50 mV/pA, dependently on the amplifier used) gains. Also use higher holding potentials (note that it might lead to membrane destabilization).
10. The diameter of the NT lumen is larger than the sizes of potassium and chloride ions (around 0.4 nm). In the absence of charge on the surface of lipid bilayer, the ion concentrations inside NT can be considered equal to that in the bulk. However, the addition of negatively charged lipids DOPS and PIP_2 to the NT membrane is required to recruit Dyn1. The surface

charge is balanced by the net charge of electrolyte inside the NT, leading to an increase of the electrolyte concentration c_{NT} in the NT lumen:

$$c_{NT} = 2c_0\sqrt{1 + \left(\frac{\alpha}{r_{NT}}\right)^2} \quad (2)$$

where $\alpha = \gamma/ec_0$, γ is the surface charge density, e is the electron charge, and c_0 is the bulk electrolyte concentration/activity. The concentration increase leads to a proportional decrease in the specific electrical resistance:

$$\rho_{NT} = \frac{2\rho_0}{\sqrt{1 + \left(\frac{\gamma}{er_{NT}c_0}\right)^2}} \quad (3)$$

where $\rho_0 = 66.6\ \Omega\cdot\text{cm}$ is the specific resistance of the bulk electrolyte. Note that at large c_0 (e.g., 1 M electrolyte concentration) $\rho_{NT} = \rho_0$. At physiological ionic strength, the measured radius increases with addition of charge lipid species; however, the introduction of Debye correction (defined by Eq. 3) fully accounts for the increase (Fig. 9).

11. Hemi-fission is a local closure of the NT lumen resulting in splitting of the NT lumen as well as the inner lipid monolayer of the NT membrane in two disconnected parts while the outer lipid monolayer remains continuous [19, 26]. Disconnection of the outer monolayer finalizes NT scission [6, 19, 26, 34]. Hemi-fission proceeds via sequential not simultaneous

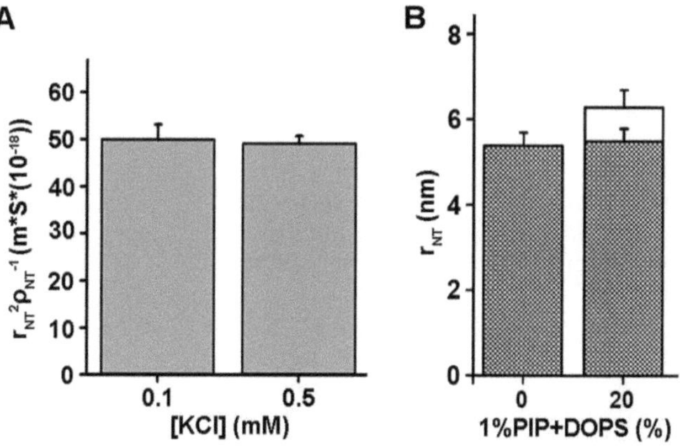

Fig. 9 Dependence of G_{NT} on charged lipid species. (**a**) Increasing the bulk electrolyte concentration does not change the corrected G_{NT} for the lipid composition used in Dyn1 experiments (From ref. 27. Reprinted with permission from AAAS). (**b**) The r_{NT} values (10 NTs analyzed in each case, the bars show SE) calculated using bulk (empty bars) and corrected (filled bars) specific resistance of the electrolyte

breaching of inner and outer lipid monolayers of the NT membrane thus avoiding pore formation. Nevertheless, coarse-grained modeling of dynamin-driven fission revealed formation of extremely short and transient pores [6]. In the following we discuss whether such pores can be detected by the conductance measurements described here. As the parameters of simulated pores cannot be directly compared with experimental observables, we estimated whether our method would detect a transient pore caused by transmembrane electric field. Field-induced pores have long been used for studying basic mechanisms of membrane poration [35, 36]. At physiological ionic strengths, the pore conductance ranges between 100 and 1000 pS and the lifetime is in the ms range [36]. In turn, the conductance of the nanotube in the pre-fission state varies from 20 to 100 pS (Fig. 7c, [26, 27]). With our detection limit of 5 pS (at 100 mV holding potential, 1 KHz bandwidth), the method shall resolve the individual lipid pores in the nanotube wall. However, if Dyn1 diminishes the pore lifetime and/or conductance, then those pores can fall below the resolution of our method. Further experimental/modeling efforts would be required to determine whether Dyn1 promote membrane poration, e.g., due to increased membrane insertion in the transition state of GTP hydrolysis [19].

Acknowledgments

The work was partially supported by the Russian Foundation for Basic Research (project # 17-04-02042) and the Spanish Ministry of Science, Innovation and Universities grants BFU2015-70552-P and PGC2018-099971-B-I00 (MCIU/AEI/FEDER, UE).

References

1. Chernomordik LV, Kozlov MM (2003) Protein-Lipid interplay in fusion and fission of biological membranes. Annu Rev Biochem 72:175–207
2. Campelo F, Arnarez C, Marrink SJ, Kozlov MM (2014) Helfrich model of membrane bending: From Gibbs theory of liquid interfaces to membranes as thick anisotropic elastic layers. Adv Colloid Interf Sci 208:25–33
3. Kozlov MM, McMahon HT, Chernomordik LV (2010) Protein-driven membrane stresses in fusion and fission. Trends Biochem Sci 35:699–706
4. Chlanda P, Mekhedov E, Waters H, Schwartz CL, Fischer ER, Ryham RJ, Cohen FS, Blank PS, Zimmerberg J (2016) The hemifusion structure induced by influenza virus haemagglutinin is determined by physical properties of the target membranes. Nat Microbiol 1:16050
5. Haldar S, Mekhedov E, McCormick CD, Blank PS, Zimmerberg J (2019) Lipid-dependence of target membrane stability during influenza viral fusion. J Cell Sci 132:jcs218321
6. Pannuzzo M, McDargh ZA, Deserno M (2018) The role of scaffold reshaping and disassembly in dynamin driven membrane fission. Elife 7:e39441
7. Shangguan T, Alford D, Bentz J (1996) Influenza virus–liposome lipid mixing is leaky and largely insensitive to the material properties of the target membrane. Biochemistry 35:4956–4965

8. Villar AV, Alonso A, Goni FM (2000) Leaky vesicle fusion induced by phosphatidylinositol-specific phospholipase C: observation of mixing of vesicular inner monolayers. Biochemistry 39:14012–14018
9. Yang ST, Zaitseva E, Chernomordik LV, Melikov K (2010) Cell-penetrating peptide induces leaky fusion of liposomes containing late endosome-specific anionic lipid. Biophys J 99:2525–2533
10. Montessuit S, Somasekharan SP, Terrones O, Lucken-Ardjomande S, Herzig S, Schwarzenbacher R, Manstein DJ, Bossy-Wetzel E, Basanez G, Meda P, Martinou JC (2010) Membrane remodeling induced by the dynamin-related protein Drp1 stimulates Bax oligomerization. Cell 142:889–901
11. Frolov VA, Lizunov VA, Dunina-Barkovskaya AY, Samsonov AV, Zimmerberg J (2003) Shape bistability of a membrane neck: A toggle switch to control vesicle content release. Proc Natl Acad Sci U S A 100:8698–8703
12. Ratinov V, Plonsky I, Zimmerberg J (1998) Fusion pore conductance: experimental approaches and theoretical algorithms. Biophys J 74:2374–2387
13. Cabeza JM, Acosta J, Ales E (2010) Dynamics and regulation of endocytotic fission pores: role of calcium and dynamin. Traffic 11:1579–1590
14. Lindau M, Rosenboom H, Nordmann J (1994) Exocytosis and endocytosis in single peptidergic nerve terminals. Adv Second Messenger Phosphoprotein Res 29:173–187
15. Spruce AE, Iwata A, White JM, Almers W (1989) Patch clamp studies of single cell-fusion events mediated by a viral fusion protein. Nature 342:555–558
16. Almers W, Breckenridge LJ, Spruce AE (1989) The mechanism of exocytosis during secretion in mast cells. Soc Gen Physiol Ser 44:269–282
17. Zimmerberg J, Blumenthal R, Sarkar DP, Curran M, Morris SJ (1994) Restricted movement of lipid and aqueous dyes through pores formed by influenza hemagglutinin during cell fusion. J Cell Biol 127:1885–1894
18. Rosenboom H, Lindau M (1994) xo-endocytosis and closing of the fission pore during endocytosis in single pituitary nerve terminals internally perfused with high calcium concentrations. Proc Natl Acad Sci U S A 91:5267–5271
19. Mattila JP, Shnyrova AV, Sundborger AC, Hortelano ER, Fuhrmans M, Neumann S, Muller M, Hinshaw JE, Schmid SL, Frolov VA (2015) A hemi-fission intermediate links two mechanistically distinct stages of membrane fission. Nature 524:109–113
20. Frolov VA, Dunina-Barkovskaya AY, Samsonov AV, Zimmerberg J (2003) Membrane permeability changes at early stages of influenza hemagglutinin-mediated fusion. Biophys J 85:1725–1733
21. Ferguson SM, De Camilli P (2012) Dynamin, a membrane-remodelling GTPase. Nat Rev Mol Cell Biol 13:75–88
22. Antonny B, Burd C, De Camilli P, Chen E, Daumke O, Faelber K, Ford M, Frolov VA, Frost A, Hinshaw JE, Kirchhausen T, Kozlov MM, Lenz M, Low HH, McMahon H, Merrifield C, Pollard TD, Robinson PJ, Roux A, Schmid S (2016) Membrane fission by dynamin: what we know and what we need to know. EMBO J 35:2270–2284
23. Danino D, Hinshaw JE (2001) Dynamin family of mechanoenzymes. Curr Opin Cell Biol 13:454–460
24. Schmid SL, Frolov VA (2011) Dynamin: functional design of a membrane fission catalyst. Annu Rev Cell Dev Biol 27:79–105
25. Ugarte-Uribe B, Garcia-Saez AJ (2017) Apoptotic foci at mitochondria: in and around Bax pores. Philos Trans R Soc Lond Ser B Biol Sci 372:20160217
26. Bashkirov PV, Akimov SA, Evseev AI, Schmid SL, Zimmerberg J, Frolov VA (2008) GTPase cycle of dynamin is coupled to membrane squeeze and release, leading to spontaneous fission. Cell 135:1276–1286
27. Shnyrova AV, Bashkirov PV, Akimov SA, Pucadyil TJ, Zimmerberg J, Schmid SL, Frolov VA (2013) Geometric catalysis of membrane fission driven by flexible dynamin rings. Science 339:1433–1436
28. Simunovic M, Prévost C, Callan-Jones A, Bassereau P (2016) Physical basis of some membrane shaping mechanisms. Philos Trans A Math Phys Eng Sci 374:20160034
29. Tunuguntla RH, Escalada A, Frolov VA, Noy A (2016) Synthesis, lipid membrane incorporation, and ion permeability testing of carbon nanotube porins. Nat Protoc 11:2029–2047
30. Neher E, Sakmann B (1995) Single-channel recording. Plenum Press, New York
31. Bashkirov PV, Frolov VA (2018) Mechanochemistry and catalysis of membrane fission: lessons from dynamins. J Phys D Appl Phys 51:343001
32. Prévost C, Tsai FC, Bassereau P, Simunovic M (2017) Pulling Membrane Nanotubes from Giant Unilamellar Vesicles. J Vis Exp e56086. https://doi.org/10.3791/56086

33. Howorka S, Siwy Z (2009) Nanopore analytics: sensing of single molecules. Chem Soc Rev 38:2360–2384
34. Zhang G, Muller M (2017) Rupturing the hemi-fission intermediate in membrane fission under tension: Reaction coordinates, kinetic pathways, and free-energy barriers. J Chem Phys 147:064906
35. Evans E, Heinrich V, Ludwig F, Rawicz W (2003) Dynamic tension spectroscopy and strength of biomembranes. Biophys J 85:2342–2350
36. Melikov KC, Frolov VA, Shcherbakov A, Samsonov AV, Chizmadzhev YA, Chernomordik LV (2001) Voltage-Induced Nonconductive Pre-Pores and Metastable Single Pores in Unmodified Planar Lipid Bilayer. Biophys J 80:1829–1836

Chapter 12

Integrating Optical and Electrochemical Approaches to Assess the Actions of Dynamin at the Fusion Pore

Katherine A. Smith, Emily R. Prantzalos, and Arun Anantharam

Abstract

Of the techniques currently available to monitor dense core granule exocytosis in adrenal chromaffin cells, two have proven particularly useful: carbon-fiber amperometry and total internal reflection fluorescence (TIRF) microscopy. Amperometry enables the detection of oxidizable catecholamines escaping a fusion pore with millisecond time resolution. TIRF microscopy, and its variant polarized-TIRF (pTIRF) microscopy, provides information on the characteristics of fusion pores at temporally later stages. Used in conjunction, amperometry and TIRF microscopy allow an investigator to follow the fate of a fusion pore from its formation to expansion or reclosure. The properties of fusion pores, including their structure and dynamics, have been shown by multiple groups to be modified by the dynamin GTPase (Dyn1). In this chapter, we describe how amperometry and TIRF microscopy enable insights into dynamin-dependent effects on exocytosis in primary cultures of bovine adrenal chromaffin cells.

Key words Dynamin, pTIRF, TIRF, Amperometry, Fusion, Fusion pore, Chromaffin, Exocytosis, Endocytosis, Fission, Granule

1 Introduction

At one time, exocytosis was conceptualized as a homogenous process that was subject to uniform regulation and which ended, inexorably, with the collapse of a fused granule into the plasma membrane. The corollary of such a view was that with each fusion event, a granule's complete lumenal content was released. As the field of exocytosis has matured and our ability to measure events associated with exocytosis has improved, it is evident that after a fusion pore forms, there is no obligation for it to expand widely or rapidly, and indeed that expansion itself may be subject to regulation [1–5].

A protein that imposes definite kinetic and structural constraints on fusion pore expansion, independent of its role in fission, is the protein dynamin [5–9]. Although the mechanism by which dynamin regulates fusion pore properties is not yet resolved, the

Rajesh Ramachandran (ed.), *Dynamin Superfamily GTPases: Methods and Protocols*, Methods in Molecular Biology, vol. 2159, https://doi.org/10.1007/978-1-0716-0676-6_12,

activity of its GTPase domain is thought to be of central importance. This is evident from studies in adrenal chromaffin cells overexpressing wild-type (WT) dynamin, or dynamin with enhanced (T141A) or diminished (T65A) GTPase activities [5]. Dynamin T141A promotes the rate of fusion pore expansion, resulting in faster rates of cargo release. The T65A mutant, on the other hand, slows pore expansion, resulting in slower rates of cargo release. Two approaches ideally suited to monitor dynamin's effects on fusion pores include carbon-fiber amperometry and TIRF microscopy. In the next two paragraphs, we will describe the broad principles by which these techniques work and how they can be applied to studies of secretion in primary bovine adrenal chromaffin cells cultured on glass-bottom culture dishes.

Amperometry uses a fixed-potential, carbon-fiber electrode (CFE) to oxidize catecholamines (e.g., norepinephrine, epinephrine, or dopamine) as they are released from cells. This reaction produces electrons and, therefore, a current which can be measured [10–14]. The magnitude of the current is proportional to the amount of the catecholamine released by the cell [14]. A "typical" amperometric event will have two components: a pre-spike foot (PSF) and a larger spike that rises and decays with rapid kinetics (Fig. 1). The PSF conveys information about the initial fusion pore, which may be so narrow as to only enable a "trickle" of outward catecholamine flux [13]. The larger spike is likely to represent the larger quantity of catecholamine released as the fusion pore expands. Its rate of rise and decay may be related to the rate at which the catecholamines are completely released [14].

Eventually, fusion pores can dilate to a geometry where dense core cargo release is possible. These later stages of expansion may not be detectable with CFE, as much of the granule catecholamine content may already have been released by this time. To monitor the kinetically slower process of dense core cargo release, TIRF microscopy is frequently used [15, 16]. The incorporation of polarization optics (pTIRF) to a standard TIRF microscope enables the additional possibility of detecting membrane curvature changes associated with fusion and release of peptides (Fig. 2) [17].

The details of pTIRF microscopy have been extensively described elsewhere [18, 19] but will be described again here in some detail. Briefly, the technique requires the use of a membrane-embedded dye with a known orientation such as 1,1′-dioctadecyl-3,3,3′,3′-tetramethylindocarbocyanine perchlorate (diI). Membranes labeled with diI are exposed to sequential p- and s-polarized TIR laser excitation. From the resulting P and S emission data, pixel-to-pixel P/S ratio images and $P + 2S$ sum images are calculated offline. Combined with computer simulations, the amplitude of the P/S can be used to predict the topology of membrane deformations occurring during exocytosis. The linear combination $P + 2S$ reports the approximate carbocyanine dye

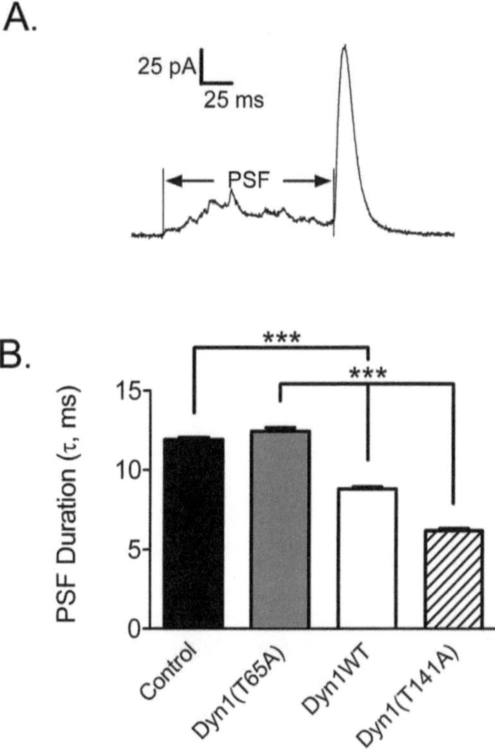

Fig. 1 Dyn1 GTPase activity regulates early fusion pore lifetime. (**a**) An exemplar amperometric trace of a stimulated chromaffin cell is shown. The amperometric current consists of two components: a long PSF and a large spike. (**b**) Analysis of PSF durations reveal that the lifetime of the early fusion pore is modified by Dyn1 in a GTPase-dependent manner (Adapted from [5])

emission at any particular region of the membrane. The $P + 2S$ parameter is theoretically proportional to the amount of dye at any location convolved with the evanescent field intensity. Thus, the value of $P + 2S$ will increase when a greater amount of fluorescently labeled membrane is close to the glass interface and decrease when the dye diffuses away from the substrate into a post-fusion membrane indentation. Together, the P/S and $P + 2S$ can be used to predict the approximate topology of a fused granule as the pore either expands, contracts, or reseals with endocytosis (Fig. 2) [17].

2 Materials

Prepare all solutions using sterile, autoclaved deionized water and analytical grade reagents. Prepare and store all chemicals and reagents according to manufacturer requirements. Store dyes away from direct light. Diligently follow all waste disposal regulations when disposing of waste materials. Whenever possible, we have listed the manufacturer/distributor of a particular reagent or piece of equipment along with a catalog number.

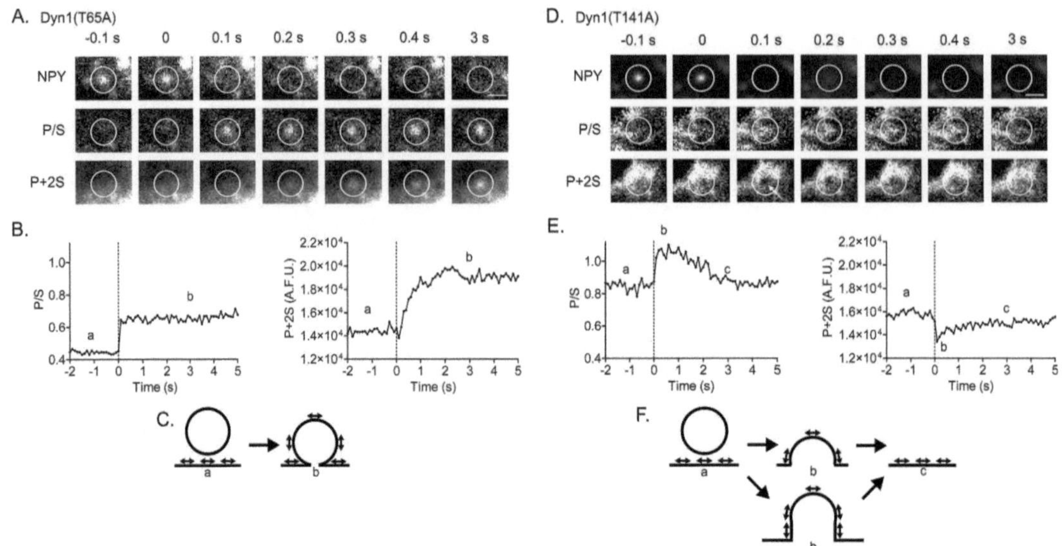

Fig. 2 Dyn1 GTPase mutants slow (Dyn1(T65A)) or hasten (Dyn1(T141A)) the widening of the fusion pore. (**a**, **d**) Chromaffin cells were co-transfected with NYP-Cer and either Dyn1(T65A) or Dyn1(T141A). Time series of NPY release and concomitant changes in membrane *P/S* and *P* + *2S* are shown. (**b**, **c**) A long-lived *P/S* change accompanied by a long-lived increase in *P* + *2S* is suggestive of a narrow pore whose expansion is slowed. (**e**, **f**) Short-lived *P/S* changes accompanied by transient *P* + *2S* changes suggest that the fusion pore is rapidly expanding, most likely resulting in the collapse or near collapse of the granule membrane into the plasma membrane. An alternative possibility (F*b*) for the fused granule/plasma membrane domain that is consistent with computer simulations is also shown (Adapted from [5])

2.1 Cell Preparation Materials

Reagents and solutions should be kept on ice unless otherwise noted.

2.1.1 Collagen Coating

1. FluoroDish Cell Culture Dish—35 mm, 23 mm well (World Precision Instruments Incorporated, Sarasota, FL, USA; catalog number FD35-100).
2. Poly-D-lysine (Advanced Biomatrix, Carlsbad, CA, USA; catalog number 5049).

2.1.2 Cell Preparation Media

1. Physiological salt solution (PSS): 145 mM NaCl, 5.6 mM KCl, and 15 mM HEPES, pH 7.4 at ~28 °C, filtered with Stericup-GP Sterile Vacuum Filtration System (MilliporeSigma, Burlington, MA, USA; catalog number: SCGPU05RE) (*see* **Note 1**).
2. Working bovine collagen solution: 0.01 N HCl solution, filter-sterilized, 3 mg/mL type 1 bovine collagen solution (Advanced BioMatrix; catalog number 5005-100ML).
3. Electroporation media: 10% fetal bovine serum (FBS) (Thermo Fisher, Waltham, MA, USA; catalog number 10438026) in DMEM/F-12 (catalog number 11320033).

4. 1× Antibiotic media: 10% FBS, 100 units/mL penicillin-streptomycin (Thermo Fisher; catalog number 15140122), 25 μg/mL gentamicin (Thermo Fisher; catalog number 15750060), and 10 μM cytosine arabinofuranoside (CAF) in DMEM/F-12.
5. 2× Antibiotic media: 10% FBS, 100 units/mL penicillin-streptomycin, and 25 μg/mL gentamicin in DMEM/F-12.
6. Fungizone media: 10% FBS, 100 units/mL penicillin-streptomycin, 25 μg/mL gentamicin, 1.3 μg/mL Amphotericine B (X-Gen Pharmaceuticals Incorporated, Horseheads, NY, USA; catalog number NDC 39822-1055-5), and 10 μM CAF in DMEM/F-12.
7. PSS-glucose: 0.56% 1 M glucose in PSS (*see* **Note 2**).

2.1.3 Cell Preparation

1. Peristaltic pump/perfusion system [16, 20].
2. String.
3. Liberase TL Research Grade (Roche, Indianapolis, IN, USA; catalog number 05040120001).
4. Liberase TH Research Grade (Roche; catalog number 5401135001).
5. Deoxyribonuclease I from bovine pancreas (DNase) (Millipore Sigma; catalog number D5025-375KU).
6. 50 mL centrifuge tubes (Corning, Corning, NY, USA).
7. Metal block.
8. Autoclaved dissection equipment (scissors, tweezers, scalpel).
9. 400 μm Nylon Mesh Filtering Screen (SEFAR NITEX; catalog number 06-400/38).
10. 250 μm Nylon Mesh Filtering Screen (SEFAR NITEX; catalog number 03-250/50).
11. 150 μm Nylon Mesh Filtering Screen (SEFAR NITEX; catalog number 03-150/38).
12. Hemocytometer (Thermo Fisher).

2.1.4 Electroporation and Transfection

1. Transfecting DNA: NPY-CER and hemagglutinin (HA)-tagged, human Dyn1 constructs as reported [5].
2. Neon Transfection System (Invitrogen (Thermo Fisher); catalog number MPK5000).
3. Neon Transfection System 100 μL Kit (Invitrogen (Thermo Fisher); catalog number MPK10025).

2.2 Amperometry Materials

2.2.1 Tool Preparation and Amperometric Measurements

1. Carbon-Fiber Electrodes (5 μm; ALA Scientific, Westbury, NY, USA).
2. Surgical knife with an unused and sterile #10 blade.
3. A standard glass slide.
4. Modeling clay.
5. Clear Scotch™ tape.
6. Amplifier: Axopatch 200A amplifier modified for extended voltage output (Axon Instruments, Foster City, CA, USA).
7. Analog to digital converter (ADC).
8. Air Table (TMC, Peabody, MA, USA).
9. Ag/AgCl reference electrode.

2.2.2 Imaging Solutions and Perfusion Components

1. Basal physiological salt solution (basal PSS): 145 mM NaCl, 5.6 mM KCl, 2.2 mM $CaCl_2$, 0.5 mM $MgCl_2$, 5.6 mM glucose, 15 mM HEPES, pH 7.4 at ~28 °C.
2. Stimulating physiological salt solution (stimulating PSS): 95 mM NaCl, 56 mM KCl, 5 mM $CaCl_2$, 0.5 mM $MgCl_2$, 5.6 mM glucose, 15 mM HEPES, pH 7.4.
3. QMM Quartz Micro Manifold (ALA Scientific Instruments; catalog number ALA QMM-4).
4. VC3 Channel Focal Perfusion System (ALA Scientific Instruments; catalog number ALA VC3X4PP).
5. Ten PSI Pressure Regulator (ALA Scientific Instruments; catalog number ALA PR10).
6. Manipulator (Thorlabs; catalog number TS 5000-150).
7. MetaMorph Imaging Software (Molecular Devices, Sunnyvale, CA, USA).

2.2.3 Data Analysis and Statistics

1. For analysis of amperometric recordings of spikes and currents, we use IGOR XOP (Wavemetrics, Portland, OR) [21].
2. Statistical analyses can be carried out using GraphPad Prism (La Jolla, CA, USA).

2.3 pTIRF Microscopy Materials

2.3.1 pTIRF Microscopy Setup Components

1. IX70 Inverted Microscope (Olympus, Center Valley, PA, USA).
2. iXon EMMCD Camera (Andor Technology, Belfast, UK).
3. 442 nm solid-state laser (CVI Melles Griot, Albuquerque, NM, USA).
4. 514 nm Argon ion laser (CVI Melles Griot).
5. Achromatic optically coated plano-concave lenses (focal length 100, 250 mm) (Thorlabs).

6. Achromatic optically coated plano-convex lenses (focal length 50, 125 mm) (Thorlabs).
7. Mounted Achromatic Quarter-Wave Plate (Thorlabs; catalog number AQWP05M-600).
8. 420–680 nm Polarizing Beamsplitter Cube (Thorlabs; catalog number PBS201).
9. Six Station Neutral Density Wheel (Thorlabs; catalog number FW1AND).
10. HQ412lp Dichroic Filter (Chroma Technology Corp, Bellows Falls, VT, USA; catalog number NC255583).
11. z442/514rpc Dichroic Filter Cube (Chroma Technology Corp).
12. z442/514 m (Chroma Technology Corp).
13. Stepper-motor Driven Smart Shutter (Sutter Instruments, Novato, CA, USA; catalog number IQ25-1219).
14. 2D (X–Y) Scanning Galvanometer Mirror System (Thorlabs; catalog number GVSM002).
15. APON 60 × 1.49 NA Objective (Olympus).

2.3.2 Imaging Solutions and Perfusion Components

Materials and solutions are identical to that of the amperometry imaging solutions and perfusion components above, in addition to the following items:

1. DiI Membrane Dye (Invitrogen (Thermo Fisher)).
2. Rhodamine 6G Chloride (Invitrogen (Thermo Fisher)).

3 Methods

3.1 Bovine Chromaffin Cell Preparation

This protocol was adapted from [20].

3.1.1 Cell Culture Dish Preparation

1. Place the desired number of 35 mm dishes (based on the number of experiments to be performed) into the biological safety cabinet and remove their lids. UV the dishes and lids for 10–15 min.
2. Add 1 mL of poly-D-lysine to each coverslip. UV the dishes and lids for 10 min.
3. Aspirate the poly-D-lysine solution from the coverslips.
4. Rinse the dishes 2–3 times with deionized water (*see* **Notes 3** and **4**).
5. Add 1 mL of the working bovine collagen solution to each dish, being careful not to allow any collagen to spill over the coverslip ridge. Leave the dishes uncovered in the biological safety cabinet overnight (*see* **Note 5**).

3.1.2 Gland Digestion and Cell Preparation

This section is to be performed on the day of bovine gland dissociation.

1. Remove the Liberase TH and TL from freezer and put on ice to thaw. Rinse perfusion tubes in a beaker of deionized water. Cut strings to a length of approximately 20 cm each. The string will be used to tie glands to perfusion system.
2. Clean the perfusion system (*see* **Note 6**). Add a small amount of the PSS-glucose solution to each chamber and allow it to drip through the perfusion line and into the chamber over which the bovine glands will be tied (*see* **Note 7**). Check that all chambers are warm.
3. While the perfusion lines are being rinsed, prepare collagenase: Add the full bottle (2 mL) of TH to 98 mL of the PSS-glucose solution and dissolve 0.0875 mg/mL DNase in the TH solution. Separate from the TH solution and add 1.15 mL TL to 84 mL of the PSS + glucose solution. Dissolve 0.074 mg/mL of DNase in the TL solution. Store the TH and TL solutions on ice. Twenty minutes prior to tissue digestion, place TH and TL solutions in the 37 °C water bath.
4. When the glands arrive, trim as much fat as possible from the glands using the sterilized scissors, scalpel, and tweezers (*see* **Note 8**). Excess fat may clog the perfusion system. Be careful not to damage the opening to the inferior phrenic artery, as this is where one end of a small plastic tube will be inserted (*see* **Note 9**).
5. Once thoroughly trimmed, insert a small plastic tube into the opening to the inferior phrenic artery of a gland. Tightly tie a string around the tissue surrounding the perfusion tube to securely affix it to the gland. The tube and string should be able to support the hanging gland without additional support.
6. Score the gland on both sides with a scalpel blade. This will help the perfusion system to distribute the collagenase to all of the capillaries of the gland and to drip out of the gland into the reservoir chamber.
7. Attach each gland to the perfusion system by inserting the plastic tube (now inserted into the phrenic artery of the bovine gland) into one of the perfusion lines (*see* **Note 10**).
8. Perfuse the glands for 5–10 min with warmed PSS-glucose (*see* **Note 11**). Aspirate any blood that has drained from the gland into the reservoir chamber and replace it with an equal amount of the warmed PSS-glucose solution. Repeat until the PSS draining out of the gland is clear (i.e., all the blood has been washed out of the gland).

9. Aspirate the PSS-glucose solution from each chamber and replace with 9 mL of the TH solution and digest the glands for 30 min.
10. On a chilled metal block, butterfly the gland open using the scissors or scalpel. Scrape the light pink medulla away from darker red cortical tissue (*see* **Note 12**). Mince the medulla.
11. Place the minced medullary tissue into a beaker containing a 2:1 ratio of TH:TL solutions and a stir bar. Gently stir the cells for 30 min at 37 °C.
12. Next, filter the liquid cell suspension through a 450 μm mesh into a flask (*see* **Note 13**).
13. Divide the filtered suspension into 2, 50 mL centrifuge tubes and spin for 5 min at 500–1000 × *g*. Remove supernatant (usually pouring the supernatant out will work), leaving only a small amount of liquid behind with the cells (*see* **Note 14**). Resuspend cells in the PSS-glucose solution and pipette up and down to mix.
14. Repeat **steps 12** and **13** with a 250 μm mesh and then a 150 μm mesh. After removing the supernatant from the final filtered product, resuspend cells in PSS-glucose and proceed to count cells.

3.1.3 Cell Counting

1. Dilute cells in small volume of PSS-glucose to make counting manageable. Determine the number of cells using a hemocytometer.
2. Centrifuge the collected cells at 500–1000 × *g*, discard the supernatant, and resuspend the cells in R buffer such that there are one million cells per 100 μL of R buffer.

3.1.4 Electroporation and Transfection

The following procedure assumes three dishes of each one of four different conditions will be electroporated: NPY-Cerulean (Cer) alone as well as NPY-Cer co-transfected with Dyn1 (WT), Dyn1 (T65A), or Dyn1 (T141A).

1. Prepare cells for electroporation: Add four million cells diluted in R buffer to each of four microfuge tubes (*see* **Note 15**).
2. Add the appropriate amount of DNA to each microfuge tube: Add a total of 15 μg of DNA per one million cells (*see* **Note 16**).
3. Prepare cells for electroporation at a final dilution of one million cells/mL of electroporation media. Pipette the contents of the microfuge tube of the appropriate construct in 100 μL increments in an electroporation pipette. Electroporate at 1100 V for 40 ms. Add the electroporated cells to the tube with electroporation media. Repeat until there is less than

100 μL of non-electroporated cells in the microfuge tube. The remaining non-electroporated cells may be discarded.

4. Repeat **step 3** for each transfection condition.
5. After each electroporation, gently pipette electroporated cells and 1 mL of antibiotic-free media into the center of each glass-bottom Fluoro Dish.
6. Repeat **step 5** for each construct.
7. Incubate dishes at 37 °C, 5% CO_2 for 4 h to allow cells to settle.
8. Four hours after plating, add 1 mL of 2× antibiotic media to each dish to bring the final concentration of antibiotic to 1× (*see* **Note 17**).
9. Incubate dishes at 37 °C, 5% CO_2 overnight.
10. The next morning, aspirate media from the plates. Add 2 mL of Fungizone Media to each plate.
11. Incubate dishes at 37 °C, 5% CO_2 overnight.
12. The next morning, aspirate media from the dishes. Add 2 mL of 1× Antibiotic Media to each plate.

3.2 Amperometry

3.2.1 Amperometric Measurements

1. Cut tip of CFE (*see* **Note 18**) [13]. Place two pieces of clay on a glass slide, with one piece of clay molded to one-half the size of the other. Apply a small piece of clear Scotch™ tape to the glass slide a short distance from the smaller piece of clay. Angle the CFE on the clay so that its tip rests on the tape. In one motion, cut the fiber tip by rolling the #10 blade over the fiber (*see* **Note 19**).
2. Insert the CFE into the electrode holder of the amplifier.
3. Immerse the CFE (held at constant potential of +650 mV) into the dish containing cells and position it so that it touches the target transfected cell.
4. Perfuse chromaffin cells with basal PSS through a needle (100 μm inner diameter) using positive pressure from a computer-controlled perfusion system (VC3 Channel Focal Perfusion System). After establishing a baseline current trace, evoke secretion via local application of the stimulating PSS for 60 s (*see* **Note 20**).
5. Collect current spike deflections using Axopatch 200A amplifier, filtered at 2 kHz and sampled at 4 kHz (*see* **Note 21**).
6. Analyze currents offline using an IGOR XOP (Wavemetrics, Portland, OR) [21].

3.3 pTIRF Microscopy

This procedure assumes the use of a 442 nm laser for excitation of cerulean-tagged NPY protein and a 514 nm laser for polarized excitation of diI. Different combinations of lasers are possible without significantly changing the setup procedure [18]. The procedure for building a pTIRF scope is described in more detail elsewhere [18, 19].

3.3.1 Implementing Polarization-Based TIRF Microscopy

1. Power on the 442 nm laser (used for conventional TIRF imaging), galvanometer mirrors, and any other necessary microscope components. Focus the laser beam onto the back focal plane (BFP) of the objective. Enter TIR by changing the position of the X-galvanometer mirror (*see* **Note 22**).
2. Using mirrors, adjust the 514 nm beam so that it is co-linear with the 442 nm beam. Position an achromatic concave lens to expand the 514 nm beam and focus it on the BFP of the 1.49 NA TIRF objective using a pair of convex lenses.
3. The 561 nm laser beam is linearly polarized as it emerges from the laser aperture and must be elliptically polarized by centering a quarter-wave (QW) plate in its path.
4. Place a polarization cube downstream of the circularly polarized beam. Polarization cubes pass light with horizontal electric field orientations and reflect the vertical component.
5. Insert a mirror into the vertically polarized beam path. To recombine the beams, position a second mirror in the horizontally polarized beam path. Place a second polarization cube where the paths meet and adjust mirrors as necessary so that beam paths are co-linear.
6. Insert motorized shutters in each of the two polarized beam paths to allow for rapid selection between p-polarized (p-pol) and s-polarized (s-pol) excitation.
7. Combine the p-pol and s-pol beams with the 442 nm beam using a dichroic mirror. The three beams should all be focused on the BFP and emerge collimated from the objective.
8. "Walk" the beam into TIR mode by changing the x position of the galvanometer mirror.

3.3.2 Imaging with pTIRF Microscopy

1. Prepare a rhodamine sample to normalize P and S emission image intensities as previously described [17] (*see* **Note 23**).
2. Prepare diI as previously described [17].
3. To stain cells with dye, directly add 10 μL of the diI solution to 2 mL of basal PSS in the culture dish. Agitate the dish gently for less than 5 s and quickly rinse the dish three times with basal PSS.
4. Place the dish containing the cells on the objective and search for one that is positive for NPY-Cer and stained with diI (*see* **Note 24**).

5. Position the tip of the local perfusion needle so it is approximately on the same focal plane as the cell of interest but just out of the field of view.
6. Using the camera software, select a region of interest for time-lapse image acquisition (*see* **Note 25**).
7. Set the perfusion to trigger with the start of image acquisition or trigger manually.
8. Start the image acquisition and shutter between the 442 nm, p-pol, and s-pol laser lines.

3.3.3 Analyzing Emission Images

1. Export images for offline analysis.
2. After correcting individual channel emission intensities using the rhodamine method (*see* **step 1** in the "Imaging with pTIRF Microscopy" section), calculate a *P/S* and *P* + *2S* image stack [17].
3. Identify *P/S* and *P* + *2S* specifically at sites of fusion by visually searching for sites of sudden NPY-Cer disappearance. A sudden disappearance of NPY-Cer indicates exocytosis has likely occurred. Commercial or free (ImageJ) software may be used to calculate emission intensities at selected ROIs (*see* **Note 26**).

4 Notes

1. To prepare five glands, at least 3, 500 mL aliquots of PSS are required. We usually make 2 L of PSS to have an extra PSS on-hand. Autoclave the filled bottles.
2. Prepare this solution on the day of cell preparation. Store at 4 °C and warm to 37 °C before use.
3. Poly-D-lysine is toxic to cells, so it is important to ensure that the dishes are thoroughly rinsed.
4. Only rinse with about 1 mL of water, being careful not to allow the water to spill over the coverslip ridge. It may also be useful to run the tip of the aspirator around the circumference of the coverslip.
5. It is essential that the hood window remains in the "up/opened" position for the night; otherwise, the water will not evaporate.
6. Allow ethanol to run through the perfusion lines (tubes) for 15 min. Drain the ethanol from the tubes. Fill the perfusion system with deionized water, turn heater on the system to the "on" position, and rinse the tubes again. Drain tubes of the water. Fill the tubes with deionized water a second time. Rinse and drain the tubes.

7. Add enough PSS-glucose solution to each chamber so that the level of the liquid is slightly below the approximate final "hung" position of each gland.
8. To assist in keeping the glands cold until perfusion, we find it useful to trim the glands on a workspace consisting of a bucket filled with ice and covered with aluminum foil.
9. When trimming the gland, leave enough of the tissue surrounding the exterior opening to the inferior phrenic artery so that the opening does not collapse. Do not use glands that lack this opening nor glands in which this opening has been cut. Glands without an intact opening may not be properly perfused.
10. We find that a good method to attach glands to the perfusion system involves gripping the gland by the plastic tube and the underside of the tissue surrounding the inferior phrenic artery with the tweezers. Doing so allows for relatively easy maneuvering of the gland into one end of the perfusion tube.
11. The PSS-glucose solution at the bottom of each chamber should have a pink or red hue (i.e., as blood drains) as the gland is perfused. If this is not the case, the perfusion system may be clogged by fat or another obstruction. For example, the gland may not have been scored enough. Alternatively, the gland may simply not have a lot of blood. Check for removable obstructions, score the gland again, and/or try squeezing the gland a few times to force the solution into the capillaries.
12. If the glands were well-perfused and tissue well-digested, the medullary portion should come off of the cortex as a gelatinous mass, with very little effort.
13. If there is a lot of undigested tissue (i.e., clumps), perform an additional incubation with TL:TH solution (2:1 ratio). Stir the cells in the beaker as before for 20 min.
14. Pippetting out the supernatant will work as well.
15. We typically use one million cells per electroporation.
16. In a typical experiment, cells might be transfected with 15 μg of NPY-Cer alone or 7.5 μg NPY-Cer with 7.5 μg of hemagglutinin (HA)-tagged, human Dyn1 plasmids. These are the constructs that were used to acquire data shown in Figs. 1 and 2, albeit using a different transfection Scheme [5].
17. Do not add the antibiotic media immediately after plating electroporated cells, as their health may be deleteriously affected.
18. Each electrode can be used in multiple experiments. Simply trim the end of the CFE using the scalpel blade to expose a new surface.

19. Test for appropriate responses by following testing procedures outlined in the ALA Scientific Carbon Fiber Electrode manual.
20. Generally, cells are perfused with PSS for 5 s and then stimulated to secrete with the elevated KCl (stimulating PSS) solution for 60 s.
21. There may be systematic differences in the sensitivity of different electrodes; the sensitivity of a single electrode may change as it is trimmed between one experiment and the next. To account for such differences, it is useful to set thresholds for spike rise time and spike amplitude. In the experiments of Fig. 2, spike amplitudes had to be greater than 10 pA and spike rise times less than 5 ms for the data to be included. Only PSFs with amplitudes greater than 1 pA and durations greater than 2.5 ms were used in PSF analysis.
22. It is often useful to use a dish of fluorescent microspheres to verify that TIR has been achieved. In epifluorescence mode, many floating microspheres will be visible. Once in TIR, the floating microspheres will disappear leaving only those within the evanescent wave plane visible.
23. The intensity of the p and s polarized lasers at the objective should be roughly equivalent. To ensure this is the case, place a dish containing rhodamine on the stage and bring it into focus. Shutter between polarized excitations and measure the emission intensity. The ratio of the resulting emission intensities (P/S) should be close 1. If not, add a neutral density filter in one polarized beam path to selectively attenuate its intensity or rotate the quarter-wave plate. In addition, one can normalize P and S emission intensities of diI to those obtained after polarized excitation of rhodamine.
24. For reasons that are still not clear to us, not every cell will take up the dye. Of course, not every cell expresses the fluorescently tagged NPY protein either. Begin the experiment by searching for transfected cells first and then verify that the transfected cell has also taken up the dye. In a cell that is stained well, there will be vivid differences in images obtained with p versus s-polarized excitation. The p-pol light will preferentially excite the diI fluorescence at the edge of a cell; the s-pol light will highlight most of the rest of the cell footprint, where the dye molecules are mostly parallel to the glass substrate.
25. Use as small an ROI as possible, usually one that is only as large as the cell of interest. Imaging with the full chip of a CCD camera will reduce acquisition speed.
26. Most aspects of image processing may be performed with Fiji (ImageJ). Computer simulations can also be performed offline (using a flexible language such as MATLAB or Interactive Data

Language (IDL)) to aid in the interpretation of membrane topological changes associated with fusion events.

Acknowledgments

We thank Noah Schenk and Dr. Mounir Bendahmane for careful reading of this manuscript. We acknowledge NIH grant GM111997 for funding support.

References

1. Perrais D, Kleppe IC, Taraska JW, Almers W (2004) Recapture after exocytosis causes differential retention of protein in granules of bovine chromaffin cells. J Physiol 560(2):413–428
2. Taraska JW, Almers W (2004) Bilayers merge even when exocytosis is transient. Proc Natl Acad Sci U S A 101:8780–8785
3. Taraska JW, Perrais D, Ohara-Imaizumi M, Nagamatsu S, Almers W (2003) Secretory granules are recaptured largely intact after stimulated exocytosis in cultured endocrine cells. Proc Natl Acad Sci U S A 100:2070–2075
4. Fulop T, Radabaugh S, Smith C (2005) Activity-dependent differential transmitter release in mouse adrenal chromaffin cells. J Neurosci 25(32):7324–7332
5. Anantharam A, Bittner MA, Aikman RL, Stuenkel EL, Schmid SL, Axelrod D, Holz RW (2011) A new role for the dynamin GTPase in the regulation of fusion pore expansion. Mol Biol Cell 22(11):1907–1918. https://doi.org/10.1091/mbc.E11-02-0101
6. Tsuboi T, McMahon HT, Rutter GA (2004) Mechanisms of dense core vesicle recapture following "kiss and run" ("cavicapture") exocytosis in insulin-secreting cells. J Biol Chem 279(45):47115–47124
7. Fulop T, Doreian B, Smith C (2008) Dynamin I plays dual roles in the activity-dependent shift in exocytic mode in mouse adrenal chromaffin cells. Arch Biochem Biophys 477(1):146–154
8. Jaiswal JK, Rivera VM, Simon SM (2009) Exocytosis of post-Golgi vesicles is regulated by components of the endocytic machinery. Cell 137(7):1308–1319. https://doi.org/10.1016/j.cell.2009.04.064
9. Shin W, Ge L, Arpino G, Villarreal SA, Hamid E, Liu H, Zhao WD, Wen PJ, Chiang HC, Wu LG (2018) Visualization of membrane pore in live cells reveals a dynamic-pore theory governing fusion and endocytosis. Cell 173(4):934–945.e912. https://doi.org/10.1016/j.cell.2018.02.062
10. Wightman RM, Schroeder TJ, Finnegan JM, Ciolkowski EL, Pihel K (1995) Time course of release of catecholamines from individual vesicles during exocytosis at adrenal medullary cells. Biophys J 68(1):383–390
11. Schroeder TJ, Jankowski JA, Kawagoe KT, Wightman RM (1992) Analysis of difusional broadening of vesicular packets of catecholamines release from biological cells during exocytosis. Anal Chem 64:3077–3083
12. Wightman RM, Jankowski JA, Kennedy RT, Kawagoe KT, Schroeder TJ, Leszczyszyn DJ, Near JA, Diliberto EJ Jr, Viveros OH (1991) Temporally resolved catecholamine spikes correspond to single vesicle release from individual chromaffin cells. Proc Natl Acad Sci U S A 88:10754–10758
13. Zhou Z, Misler S, Chow RH (1996) Rapid fluctuations in transmitter release from single vesicles in bovine adrenal chromaffin cells. Biophys J 70(3):1543–1552. https://doi.org/10.1016/S0006-3495(96)79718-7
14. Chow RH, von Ruden L, Neher E (1992) Delay in vesicle fusion revealed by electrochemical monitoring of single secretory events in adrenal chromaffin cells. Nature 356 (6364):60–63. https://doi.org/10.1038/356060a0
15. Steyer JA, Horstman H, Almers W (1997) Transport, docking and exocytosis of single secretory granules in live chromaffin cells. Nature 388:474–478
16. Allersma MW, Wang L, Axelrod D, Holz RW (2004) Visualization of regulated exocytosis with a granule-membrane probe using total internal reflection microscopy. Mol Biol Cell 15:4658–4668
17. Anantharam A, Onoa B, Edwards RH, Holz RW, Axelrod D (2010) Localized topological changes of the plasma membrane upon exocytosis visualized by polarized TIRFM. J Cell Biol 188(3):415–428. https://doi.org/10.1083/jcb.200908010

18. Passmore DR, Rao T, Anantharam A (2014) Real-time investigation of plasma membrane deformation and fusion pore expansion using polarized total internal reflection fluorescence microscopy. Methods Mol Biol 1174:263–273. https://doi.org/10.1007/978-1-4939-0944-5_18
19. Passmore DR, Rao TC, Peleman AR, Anantharam A (2014) Imaging plasma membrane deformations with pTIRFM. J Vis Exp 86. https://doi.org/10.3791/51334
20. Wick PW, Senter RA, Parsels LA, Holz RW (1993) Transient transfection studies of secretion in bovine chromaffin cells and PC12 cells: generation of kainate-sensitive chromaffin cells. J Biol Chem 268:10983–10989
21. Mosharov EV, Sulzer D (2005) Analysis of exocytotic events recorded by amperometry. Nat Methods 2(9):651–658. https://doi.org/10.1038/nmeth782
22. Anantharam A, Axelrod D, Holz RW (2012) Real-time imaging of plasma membrane deformations reveals pre-fusion membrane curvature changes and a role for dynamin in the regulation of fusion pore expansion. J Neurochem 122(4):661–671. https://doi.org/10.1111/j.1471-4159.2012.07816.x

Chapter 13

Cellular Assays for Measuring Dynamin Activity in Muscle Cells

Jessica Laiman and Ya-Wen Liu

Abstract

Dynamin is one of the best-studied membrane fission machineries, which mediates endocytic vesicle pinch-off from the plasma membrane. Among the three dynamin isoforms encoded in mammalian genome, dynamin-2 is the ubiquitously expressed isoform and leads to human muscular or neuronal diseases when mutants causing hyperactivity or hypoactivity of its membrane fission activity occur. While transferrin uptake is the most commonly used assay to measure dynamin activity in cultured cells, here we provide two different methods to quantitatively examine the activity of dynamin in myoblasts and myotubes, i.e., Bin1-tubule vesiculation and glucose transporter 4 fractionation assays, respectively. These methods could provide a quantitative measurement of dynamin activity in both differentiated and undifferentiated myoblasts.

Key words Dynamin-2, Bin1, T-Tubule, GLUT4, Muscle disorder

1 Introduction

Dynamin is a mechanochemical enzyme critical for several endocytosis pathways in different cell types [1–3]. To measure dynamin activity, the most commonly used method is the transferrin uptake assay, which works very well in many cultured cells with impaired dynamin activity. However, the transferrin uptake assay may not reflect the hyperactivity of dynamin resulting from human centronuclear myopathy-associated dynamin-2 (Dnm2) mutations (A618T and S619L mutants in Fig. 1, also *see* [4–7]). Furthermore, given the huge size of myotubes, it would be difficult to quantify the amount of internalized transferrin within a single myotube by fluorescence microscopy.

Here, we present two approaches for quantitatively analyzing dynamin activity in myoblasts and myotubes. To measure the hyperactivity of Dnm2 mutants, we took advantage of the finding that Dnm2 could mediate membrane fission of tubules coated with Bin1, a muscle-specific membrane-tubulating protein [4, 8]. After

Rajesh Ramachandran (ed.), *Dynamin Superfamily GTPases: Methods and Protocols*, Methods in Molecular Biology, vol. 2159, https://doi.org/10.1007/978-1-0716-0676-6_13, © Springer Science+Business Media, LLC, part of Springer Nature 2020

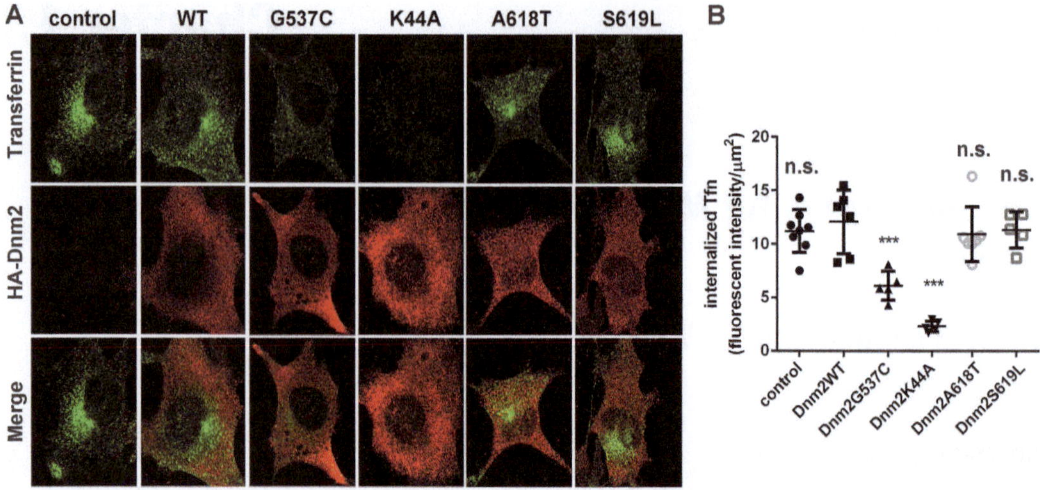

Fig. 1 Transferrin uptake assay. C2C12 myoblast ectopically expressing different HA-Dnm2 wild type (WT) or mutants was incubated with Alexa fluor 488-transferrin for 10 min. After acid wash, HA-Dnm2 was stained, and images were acquired with confocal microscopy and shown as maximum intensity projection (**a**). Intensity of transferrin and cell area was quantified with ImageJ (**b**). Bar, 10 μm. Data were compared with Dnm2WT expressing cells and analyzed with *t*-test; $***p < 0.001$

co-expression of Bin1-GFP and Dnm2-mCherry in myoblasts, the morphology of Bin1-GFP tubulation or vesiculation likely reflects the alteration of the membrane fission activity of Dnm2.

On the other hand, it has been demonstrated by several research groups that Dnm2 is essential for glucose transporter 4 (GLUT4) endocytosis in myotubes [9–11]. Therefore, to quantitatively access Dnm2 activity in myotubes, we utilize subcellular fractionation to analyze the distribution of endogenous GLUT4 in mouse C2C12-derived myotubes, which is mainly in the plasma membrane as well as in storage vesicles [12]. The ratio of GLUT4 distribution on plasma membrane versus storage vesicles reflects the activity of Dnm2 in myotubes.

2 Materials

2.1 Bin1-Tubule Vesiculation Assay in C2C12 Myoblast

2.1.1 C2C12 Culture

1. Growth medium composed of high glucose (4.5 mg/mL) DMEM supplemented with L-glutamine, sodium pyruvate, antibiotics, and 10% fetal bovine serum.
2. Trypsin-EDTA (Thermo Fisher Scientific, 15400054).
3. Cover slip: 22 mm × 22 mm, 1.5 H (*see* **Note 1**).

2.1.2 Transfection

1. Lipofectamine® 2000 (Thermo Fisher Scientific, 166-8019).
2. Opti-MEM (Thermo Fisher Scientific, 31985-070).

3. Antibiotic-free growth medium.
4. Bin1-GFP: pEGFPC1-muscle amphiphysin 2, Addgene plasmid 22213.
5. Dnm2-mCherry: Dyn2-pmCherryN1, Addgene plasmid 276989.
6. Dnm2-mCherry mutants: G537C, K44A, A618T, S619L [4].

2.1.3 Fixation and Image Acquirement

1. 4% Formaldehyde: 10% formaldehyde (Polysciences, Inc., 04018-1) was diluted with PBS into 4% (v/v) solution.
2. 0.1% Saponin: Saponin (Sigma, S7900) was dissolved in PBS to make 0.1% (w/v) solution.
3. PBS: 137 mM NaCl, 2.7 mM KCl, 10 mM Na_2HPO_4, 1.8 mM KH_2PO_4, pH 7.4.
4. Mounting solution (Southern Biotech, 0100-20).

2.2 GLUT4 Fractionation in C2C12-Derived Myotubes

2.2.1 C212 Culture and Dynamin Activity Regulation

1. Growth medium (GM) (*see* Subheading 2.1.1).
2. Differentiation medium (DM): High glucose (4.5 mg/mL) DMEM supplemented with sodium pyruvate, antibiotics and 2% horse serum, store at 4 °C.
3. 100 mm cell culture dishes.
4. Dynasore: 100 mM stock (DMSO), store at −20 °C.
5. MiTMAB: 100 mM stock (H_2O), store at −20 °C.

2.2.2 Subcellular Fractionation (See **Note 2**)

1. PBS, store at 4 °C.
2. HES: 255 mM sucrose, 20 mM HEPES (pH 7.4), 1 mM EDTA, store at 4 °C.
3. Protease inhibitor cocktail (PI): Prepared as 50× stock by dissolving one cOmplete™ Protease Inhibitor Cocktail (Roche) tablet in 1 mL autoclaved Milli-Q water. Store at −20 °C. It can be thawed and refrozen (*see* **Note 3**).
4. HES-PI: HES, 2× PI. Prepare fresh, and store at 4 °C.
5. Cell scrapers.
6. 1.7 mL microtubes.
7. 1 mL disposable syringes with 26-G needles.
8. 1× Sample buffer: 50 mM Tris (pH 6.8), 2% SDS, 10% glycerol, 0.025% bromophenol blue, 1% β-mercaptoethanol.
9. 100% Trichloroacetic acid (TCA) solution. Store at 4 °C.
10. Sterilized Milli-Q water.
11. 2 M Tris, pH 8.8.

2.2.3 Western Blot

1. GLUT4 antibody (e.g., #2213, 1:1000, Cell Signaling 1F8).
2. Na^+/K^+-ATPase α1 antibody (e.g., sc-21712, 1:1000, Santa Cruz C464.6).
3. EEA1 antibody (e.g., #3288, 1:2000, Cell Signaling C45B10).
4. α-Tubulin antibody (e.g., T6074, 1:5000, Sigma-Aldrich).

3 Methods

3.1 Bin1-Tubule Vesiculation Assay in C2C12 Myoblast

3.1.1 C2C12 Myoblast Preparation and Transfection

1. Culture mouse-derived C2C12 myoblasts (American Type Culture Collection, CRL-1772) in growth medium (GM) at 37 °C, 5% CO_2 incubator (*see* **Note 4**).
2. For transfection, seed 100,000 cells in a 35 mm dish with GM.
3. After overnight culture, mix 2.5 μg DNA (1.5 μg Bin1-GFP with 1 μg Dnm2-mCherry) with 9 μl of lipofectamine in 300 μl of Opti-MEM, and let it sit for 20 min at room temperature.
4. After changing the GM of cells into antibiotic-free growth medium, add DNA–liposome complexes evenly into the medium.
5. After 6-h incubation at 37 °C, trypsinize and subculture the cells onto 22 mm × 22 mm coverslip at 1:3 dilution in GM.
6. After overnight incubation at 37 °C, the cells are ready for following fixation and image acquirement processes.

3.1.2 Fixation and Image Acquirement

1. Twenty-four hours after transfection, fix C2C12 with 4% formaldehyde for 30 min at room temperature (*see* **Note 5**).
2. After PBS wash, permeabilize cells with 0.1% saponin for 15 min at room temperature and then wash with PBS.
3. Finally, mount the coverslips and acquire images with confocal microscope LSM700 (Fig. 2a). For each condition, capture 30 cells with comparable intensity of Bin1-GFP and Dnm2-mCherry with z-stack confocal microscopy and subsequently show and quantify the maximum intensity projected images.
4. For quantification, divide Bin1-GFP morphologies into three categories: tubular, intermediate, and vesicular, as illustrated in Fig. 2b. Tubular phenotype is equipped with more Bin1-GFP tubules, and vesicular phenotype is dominated by Bin-GFP punctate, whereas intermediate one has comparable amount of tubules and vesicles (*see* **Note 6**).
5. With this assay, Dnm2 with impaired fission activity (G537C and K44A) shows significantly higher Bin1-GFP tubule population whereas hyperactive mutants (A618T and S619L) result in dramatic vesiculation of Bin1-GFP (Fig. 2b).

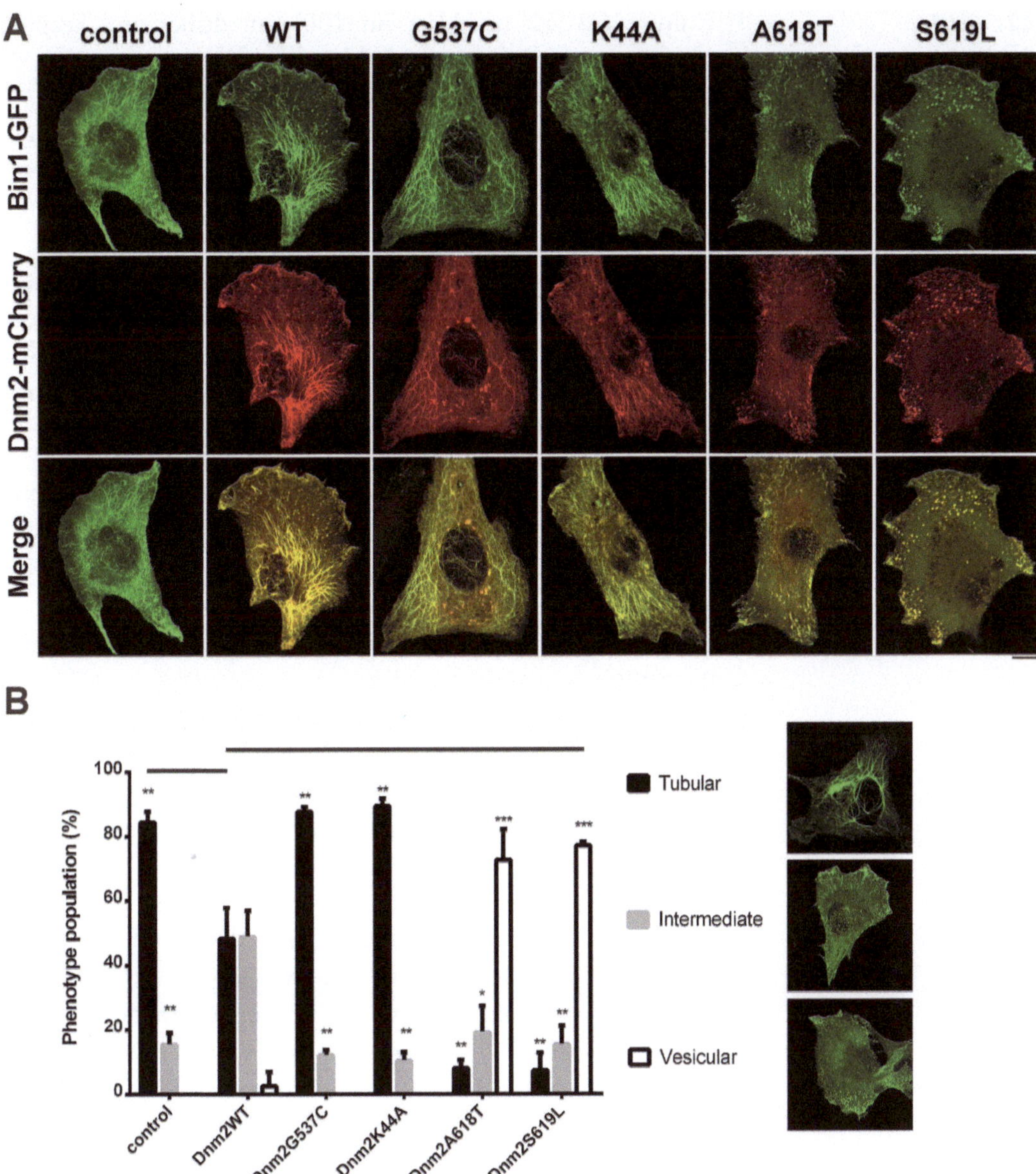

Fig. 2 Bin1-GFP tubule morphology. C2C12 myoblast ectopically expressing Bin1-GFP and different Dnm2-mCherry wild type (WT) or mutants was fixed and imaged with z-stack confocal microscopy. Maximum intensity projection images were shown (**a**) and categorized into three populations (**b**). Bar, 10 μm. Data were compared with Dnm2WT expressing cells and analyzed with *t*-test; $**p < 0.01$; $***p < 0.001$

3.2 GLUT4 Fractionation in C2C12-Derived Myotubes

3.2.1 C2C12 Culture and Dynamin Activity Regulation

1. Culture C2C12 myoblasts in 100 mm dishes and keep in condition as mentioned on Subheading 3.1.1.
2. Wait for 1–2 days until cells grow to confluence (*see* **Note 7**), and then induce cell differentiation by switching medium to DM.
3. Replace DM with fresh medium on the third or fourth day (*see* **Note 8**).
4. On the fifth day of differentiation, prepare dynamin inhibitors by diluting them (Dynasore and MiTMAB, final concentration: 80 and 5 μM, respectively) separately in pre-warmed DMEM.
5. Remove and discard DM from cultures, and gently add dynamin inhibitor solutions. Incubate at 37 °C for 30 min.

3.2.2 Subcellular Fractionation (See **Note 2**)

1. Remove and discard the medium, place the dishes on ice, and wash two times with PBS.
2. Rinse cells once with HES, and add 1250 μL HES-PI to each dish.
3. Scrape cells with cell scraper and then transfer to 1.7 mL microtube.
4. Homogenize cells by passing them through 26-G needle attached to 1 mL syringe repeatedly for 20 times.
5. Spin cell lysate at 1000 × *g* for 5 min at 4 °C to remove nuclei, and transfer supernatant to new microtube (*see* **Note 9**).
6. Spin supernatant at 5,000 × *g* for 10 min at 4 °C. After spin, transfer supernatant to new microtube; resuspend pellet in 160 μL 1× sample buffer, label as P5, and keep on ice.
7. Spin supernatant at 10,000 × *g* for 10 min at 4 °C. After spin, transfer supernatant to new microtube; resuspend pellet in 160 μL 1× sample buffer, label as P10, and keep on ice.
8. Spin supernatant at 16,000 × *g* for 20 min at 4 °C. After spin, transfer supernatant to new microtube; resuspend pellet in 160 μL 1× sample buffer, label as P16, and keep on ice (*see* **Note 10**).
9. Add 100% TCA to supernatant till the final dilution is 10%, mix well, and incubate at 4 °C for 1 h.
10. Spin at 13,000 rpm (16,000 × *g*) for 30 min at 4 °C.
11. Discard supernatant. Wash with 1 mL ice-cold sterilized Milli-Q water. Spin at 13,000 rpm (16,000 x *g*) for 15 min at 4 °C.
12. Discard water. Dissolve pellet in 160 μL 1× sample buffer, and label as S16 (*see* **Note 11**). Add Tris–HCl pH 8.8 (±2 μL) to neutralize the solution if it turns yellow until it turns back into blue.
13. Heat all samples at 60 °C for 15 min. Store at −80 °C (*see* **Note 12**).

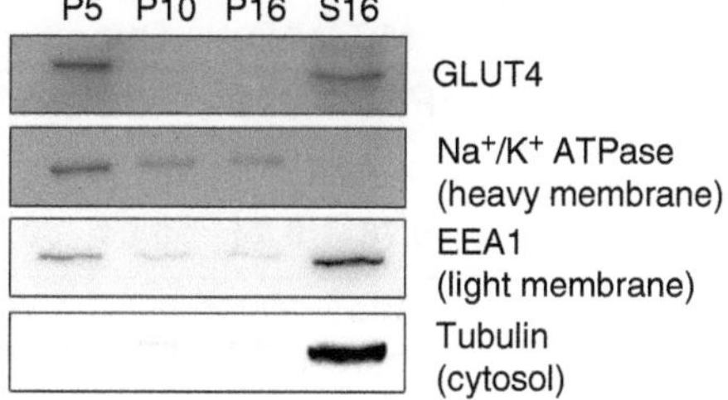

Fig. 3 GLUT4 subcellular fractionation. C2C12 myotubes were homogenized and subjected to differential centrifugation. Cells were separated into four fractions and analyzed by Western blotting for GLUT4 and several subcellular markers, including Na^+/K^+ ATPase (plasma membrane), EEA1 (early endosome), and tubulin (cytosol). The sample of 10 μL was loaded in each lane. Most of the GLUT4 is found at P5 and S16 fractions

3.2.3 Western Blot

1. Load 10 μL of each fraction on a standard SDS-PAGE gel (*see* **Note 13**) and run the gel.
2. Transfer protein to PVDF membrane by wet electrotransfer.
3. Block the membrane and then probe with anti-GLUT4 and other fraction marker antibodies with recommended dilution (*see* Subheading 2). Detect protein by using chemiluminescent HRP substrate (*see* **Note 14**).
4. Using this approach, we found endogenous GLUT4 distributes mainly in the P5 and S16 in C2C12 myotube, which is consistent with its localization in plasma membrane and storage vesicles (Fig. 3 [10, 12]). Upon 30 min of dynamin activity inhibition, the ratio of GLUT4 in heavy membrane fraction P16 increases by about 20% (Fig. 4).

4 Notes

1. Coverslips were washed with 1 N HCl and subsequently with water and ethanol before usage to help cells stick to glass.
2. All reagents and materials for GLUT4 fractionation, including microtubes, syringes, etc., must be precooled at 4 °C and kept on ice throughout the experiment to minimize protein degradation.
3. It is recommended to use the 50× stock within one month after preparation to ensure maximum efficiency.
4. C2C12 must not be allowed to become confluent as this will deplete the myoblastic population in the culture.
5. For Bin1-tubule vesiculation assay, unless necessary, all incubation should be done in darkness.

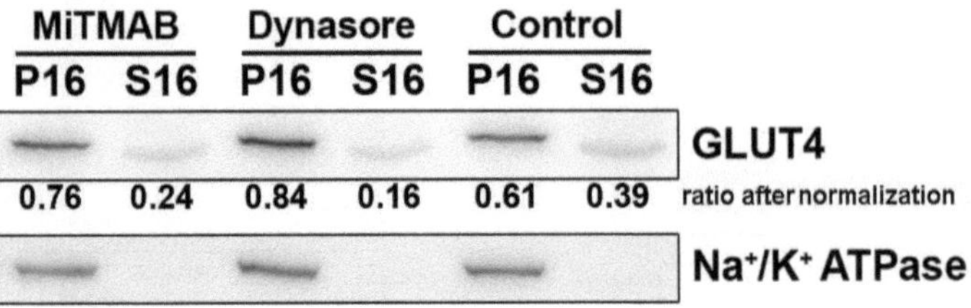

Fig. 4 GLUT4 subcellular fractionation under dynamin inhibition. C2C12 myotubes were treated with dynamin inhibitors MiTMAB or Dynasore, for 30 min, homogenized and subjected to differential centrifugation. Cells were separated into two fractions and analyzed by Western blotting for GLUT4 and Na^+/K^+ ATPase (plasma membrane marker). The amount of loaded P16 sample was two times of S16. The numbers below GLUT4 protein bands refer to GLUT4 localization ratio after normalization measured using ImageJ. Note that the distribution of GLUT4 in S16 decreased after dynamin inhibitors treatment

6. Three independent experiments should be performed in order to have statistical analysis.
7. It is best to wait until C2C12 myoblasts reach 90–95% confluency before switching to differentiation medium in order to yield more myotubes.
8. Check cell morphology under microscope every day, you should be able to observe myotubes from the third day, and they will continue to grow until the fifth day.
9. For more meticulous result, you could repeat this step once more. This optional step could also be applied to the centrifugation that follows. If you choose to do so, combine the pellet in total volume of 160 μL 1× sample buffer for each speed fraction.
10. Based on Fig. 2, GLUT4 is mainly located at P5 and S16 fractions, so we bypass **steps 6** and **7** and only use 16,000 × *g* centrifugation speed to separate the sample into P16 and S16 as in Fig. 4.
11. The pellet could be quite hard to dissolve. Make sure to keep vortexing the pellet until it is completely dissolved in the sample buffer; this step might take some time.
12. Avoid boiling the sample as high temperature might cause aggregation in some membrane proteins. It is best to run the sample as soon as possible since GLUT4 tends to be degraded quite rapidly (within a week) even when it is stored in −80 °C.
13. Both 10% and 12.5% acrylamide could be used for this experiment. However, 12.5% gel has better resolution for GLUT4 based on our experience.
14. We recommend using HRP substrate with higher sensitivity (e.g., Luminata™ Forte by Millipore) for GLUT4 because its signal is relatively weak.

Acknowledgments

This work was supported by Ministry of Science and Technology (MOST) grant 104-2320-B-002-061-MY3 and National Taiwan University grant NTU-CDP-105R7878 to Y.-W. Liu.

References

1. Schmid SL, Frolov VA (2011) Dynamin: functional design of a membrane fission catalyst. Annu Rev Cell Dev Biol 27:79–105. https://doi.org/10.1146/annurev-cellbio-100109-104016
2. Ferguson SM, De Camilli P (2012) Dynamin, a membrane-remodelling GTPase. Nat Rev Mol Cell Biol 13(2):75–88. https://doi.org/10.1038/nrm3266
3. Antonny B, Burd C, De Camilli P, Chen E, Daumke O, Faelber K, Ford M, Frolov VA, Frost A, Hinshaw JE, Kirchhausen T, Kozlov MM, Lenz M, Low HH, McMahon H, Merrifield C, Pollard TD, Robinson PJ, Roux A, Schmid S (2016) Membrane fission by dynamin: what we know and what we need to know. EMBO J 35(21):2270–2284. https://doi.org/10.15252/embj.201694613
4. Chin YH, Lee A, Kan HW, Laiman J, Chuang MC, Hsieh ST, Liu YW (2015) Dynamin-2-mutations associated with centronuclear myopathy are hypermorphic and lead to T-tubule fragmentation. Hum Mol Genet 24(19):5542–5554. https://doi.org/10.1093/hmg/ddv285
5. Bitoun M, Durieux AC, Prudhon B, Bevilacqua JA, Herledan A, Sakanyan V, Urtizberea A, Cartier L, Romero NB, Guicheney P (2009) Dynamin 2 mutations associated with human diseases impair clathrin-mediated receptor endocytosis. Hum Mutat 30(10):1419–1427. https://doi.org/10.1002/humu.21086
6. Kenniston JA, Lemmon MA (2010) Dynamin GTPase regulation is altered by PH domain mutations found in centronuclear myopathy patients. EMBO J 29(18):3054–3067. https://doi.org/10.1038/emboj.2010.187
7. Liu YW, Lukiyanchuk V, Schmid SL (2011) Common membrane trafficking defects of disease-associated dynamin 2 mutations. Traffic 12(11):1620–1633. https://doi.org/10.1111/j.1600-0854.2011.01250.x
8. Lee E, Marcucci M, Daniell L, Pypaert M, Weisz OA, Ochoa GC, Farsad K, Wenk MR, De Camilli P (2002) Amphiphysin 2 (Bin1) and T-tubule biogenesis in muscle. Science 297(5584):1193–1196. https://doi.org/10.1126/science.1071362
9. Antonescu CN, Foti M, Sauvonnet N, Klip A (2009) Ready, set, internalize: mechanisms and regulation of GLUT4 endocytosis. Biosci Rep 29(1):1–11. https://doi.org/10.1042/BSR20080105
10. Antonescu CN, Diaz M, Femia G, Planas JV, Klip A (2008) Clathrin-dependent and independent endocytosis of glucose transporter 4 (GLUT4) in myoblasts: regulation by mitochondrial uncoupling. Traffic 9(7):1173–1190. https://doi.org/10.1111/j.1600-0854.2008.00755.x
11. Hartig SM, Ishikura S, Hicklen RS, Feng Y, Blanchard EG, Voelker KA, Pichot CS, Grange RW, Raphael RM, Klip A, Corey SJ (2009) The F-BAR protein CIP4 promotes GLUT4 endocytosis through bidirectional interactions with N-WASp and dynamin-2. J Cell Sci 122(Pt 13):2283–2291. https://doi.org/10.1242/jcs.041343
12. Niu W, Bilan PJ, Ishikura S, Schertzer JD, Contreras-Ferrat A, Fu Z, Liu J, Boguslavsky S, Foley KP, Liu Z, Li J, Chu G, Panakkezhum T, Lopaschuk GD, Lavandero S, Yao Z, Klip A (2010) Contraction-related stimuli regulate GLUT4 traffic in C2C12-GLUT4myc skeletal muscle cells. Am J Physiol Endocrinol Metab 298(5):E1058–E1071. https://doi.org/10.1152/ajpendo.00773.2009

Chapter 14

Measuring Drp1 Activity in Mitochondrial Fission In Vivo

Di Hu and Xin Qi

Abstract

Mitochondrial fission is mainly regulated by a number of dynamin superfamily proteins or DSPs, of which dynamin-like protein 1 (Drp1) is responsible for the scission process during mitochondrial fission. Here we describe several methods, including monitoring mitochondrial distribution, phosphorylation, and tetramer level of Drp1, to examine the activity of Drp1 in mitochondrial fission in vivo.

Key words Mitochondrial dynamics, Dynamin-related protein 1, Phosphorylation, Western blotting, Tetramerization

1 Introduction

Mitochondrial dynamics, a balanced process of fusion and fission, is one of the major quality control strategies to maintain mitochondrial health [1, 2]. Mitochondrial dynamics is mainly regulated by a number of GTPase family proteins. While MFN1/2 and OPA1 are responsible for the fusion process, dynamin-like protein 1 (Drp1) regulates mitochondrial fission along with its adaptors proteins such as Fis1 and Mff [3, 4]. Drp1 is located mainly in the cytosol. Upon activation, Drp1 oligomerizes and is recruited to punctate spots on the mitochondrial surface [5]. Drp1 binds to mitochondrial adaptors, assembles future fission sites, and severs mitochondrial membrane in a GTP hydrolysis-dependent manner [6]. The activation of Drp1 is regulated by posttranslational modification, among which phosphorylation of Drp1 can either inhibit or activate its enzymatic activity. Several Drp1 phosphor-sites have been identified, of which ser616 phosphorylation that enhances Drp1 function and ser637 phosphorylation that inhibits Drp1 activity were well demonstrated [7–9]. As Drp1 oligomerization and translocation to mitochondria is required for mitochondrial fission, Drp1 activity can be examined by measuring the level of mitochondrial Drp1 and Drp1 tetramer species. Besides, though not the case in all, the level of phosphorylated Drp1 (S616/S637) can provide

Rajesh Ramachandran (ed.), *Dynamin Superfamily GTPases: Methods and Protocols*, Methods in Molecular Biology, vol. 2159, https://doi.org/10.1007/978-1-0716-0676-6_14,

clues on Drp1 activity. Here, based on our previous studies focusing on regulating mitochondrial translocation of Drp1 by specific inhibitors, we describe several methods, including examining mitochondrial Drp1, Drp1 tetramer, and Drp1 phosphorylation (S616/S637), to measure the activity of Drp1 in mitochondrial fission.

2 Materials

Prepare all solutions using double-distilled water and analytical grade reagents. Prepare and store all reagents at room temperature (unless indicated otherwise). Follow waste disposal regulations when disposing of waste materials.

2.1 Preparation of Total Cell Lysis

1. Phosphate buffered saline (PBS, pH 7.4) (*see* **Note 1**).
2. Total cell lysis buffer: 50 mM Tris–HCl, pH 7.5, 150 mM NaCl, 1% Triton X-100, and protease inhibitor cocktail and phosphatase inhibitor cocktail (*see* **Note 2**).

2.2 Isolation of Mitochondrial Fraction

1. Phosphate buffered saline (PBS, pH 7.4) (*see* **Note 1**).
2. Mitochondrial lysis buffer: 250 mM sucrose, 20 mM HEPES-NaOH, pH 7.5, 10 mM KCl, 1.5 mM $MgCl_2$, 1 mM EDTA, protease inhibitor cocktail, and phosphatase inhibitor cocktail (*see* **Note 2**).

2.3 Tris–Glycine SDS-Polyacrylamide Gel

1. 5% Stacking gel.
2. 10% Resolving gel.
3. SDS-PAGE running buffer.
4. 2× sample loading buffer (nonreducing): Add 5 mL 1 M Tris (pH 7), 25 mL 20% SDS, 20 mL glycerol, 2 mg bromophenol blue, and water up to 100 mL.
5. 2× sample loading buffer (reducing): 950 μL 2× nonreducing sample loading buffer with 50 μL β-mercaptoethanol. Store at −20 °C (*see* **Note 3**).

2.4 Western Blot

1. Nitrocellulose membranes.
2. Transfer buffer.
3. Tris-buffered saline (TBS).
4. TBS containing 0.05% Tween-20 (TBST).
5. Blocking solution: 5% milk in TBS. Store at 4 °C.
6. ECL solution: Add 12.2 μL H_2O_2 (10 M), 200 μL luminol (250 mM), 80 μL coumaric acid (90 mM), 4 mL Tris–HCl (1 M, pH 8.5), and water up to 40 mL. Store at 4 °C (*see* **Note 4**).

2.5 Antibodies

1. Mouse anti-DLP1 antibody (BD bioscience Cat No. 611113).
2. Phospho-DRP1 (Ser616) antibody (Cell Signaling Cat No. 3455).
3. Phospho-DRP1 (Ser637) antibody (Cell Signaling Cat No. 4867).
4. Anti-VDAC1/Porin antibody (Abcam Cat No. ab34726).
5. Anti-β-Actin antibody (Sigma-Aldrich Cat No. A1978).
6. Goat anti-mouse HRP (Thermo Scientific/Fisher, Cat No. 31430).
7. Goat anti-rabbit HRP (Thermo Scientific/Fisher, Cat No. 31460).

3 Methods

Carry out all procedures at room temperature unless otherwise specified. Use 5% stacking gel and 10% resolving gel in all of the SDS-PAGE as indicated below.

3.1 Examine Mitochondrial Drp1

1. Gently wash the cell with cold PBS twice (*see* **Note 5**).
2. Incubate the cell on ice for 30 min in mitochondrial lysis buffer.
3. Collect the cell; homogenize/disrupt the cell 20 times by repeated aspiration through a 25-G needle, followed by a 30-G needle.
4. Spin the homogenates at 800 × *g* for 10 min at 4 °C.
5. Collect the supernatant and spin at 10,000 × *g* for 20 min at 4 °C.
6. Keep the supernatant as cytosol fraction.
7. Wash the pellet with mitochondrial lysis buffer. Spin again at 10,000 × *g* for 20 min at 4 °C.
8. Keep the pellet as mitochondrial fraction (*see* **Note 6**).
9. Suspend the pellet in mitochondrial lysis buffer containing 1% Triton-X-100.
10. Measure the protein concentration of the mitochondrial fraction and prepare equal amount of proteins for load in each group.
11. Prepare the sample using 1× reducing sample loading buffer and boil at 100 °C for 10 min.
12. Run the SDS-PAGE.
13. Transfer the protein to nitrocellulose membrane.
14. Rinse the membrane once with TBST.
15. Block the membrane with blocking solution for 1 h.

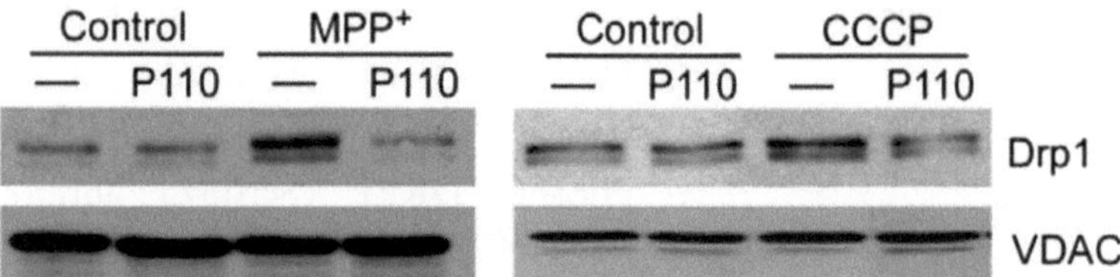

Fig. 1 Mitochondrial stressors induce Drp1 translocation and are inhibited by inhibitor P110. Cultured human SH-SY5Y neuronal cells were treated with peptide P110 prior to incubation in the presence or absence of mitochondrial stressors MMP+ or CCCP (Reproduced from Ref. 10)

16. Cut the membrane into half at the place where the standard maker represents 50 kDa.
17. Incubate the top-half membrane with mouse anti-DLP1 antibody diluted in blocking solution (1:1000) and the bottom-half membrane with mouse anti-VDAC diluted in blocking solution (1:1000) at 4 °C overnight.
18. Wash the membranes with TBST for 5 min and totally four times.
19. Incubate the membranes with goat anti-mouse HRP diluted in blocking solution (1:5000) for 1 h.
20. Wash the membrane with TBST for 5 min and totally four times.
21. Incubate the membrane with ECL solution for 2 min.
22. Develop the membrane.
23. The band shown around 75 kDa is Drp1, and the band shown around 37 kDa is VDAC. The relative level of mitochondrial Drp1 is determined as the ratio of Drp1 density over VDAC density (Fig. 1) (*see* **Note 7**).

3.2 Examine Drp1 Phosphorylation

1. Gently wash the cell with cold PBS twice (*see* **Note 5**).
2. Incubate the cell on ice for 30 min in total cell lysis buffer.
3. Collect the cell and spin for 10 min at 12,000 rpm (or 13,523 × *g*) at 4 °C.
4. Keep the supernatant as total cell lysates (*see* **Note 8**).
5. Measure the protein concentration of the total cell lysates and prepare equal amount of proteins for load in each group.
6. Prepare the sample using 1× reducing sample loading buffer and boil at 100 °C for 10 min.

7. Run the SDS-PAGE.
8. Transfer the protein to nitrocellulose membrane.
9. Rinse the membrane once with TBST.
10. Block the membrane with blocking solution for 1 h.
11. Cut the membrane into half at the place where the standard maker represents 50 kDa.
12. Incubate the top-half membrane with phospho-DRP1 (Ser616) or phosphor-DRP1 (Ser637) antibody diluted in blocking solution (1:1000) and the bottom-half membrane with mouse anti-β-actin diluted in blocking solution (1:1000) at 4 °C overnight.
13. Wash the membranes with TBST for 5 min and totally four times.
14. Incubate the membranes with the corresponding goat anti-HRP diluted in blocking solution (1:5000) for 1 h.
15. Wash the membrane with TBST for 5 min and totally four times.
16. Incubate the membrane with ECL solution for 2 min.
17. Develop the membrane.
18. The band shown around 75 kDa is phosphorylated Drp1 (S616/637) and the band shown around 37 kDa is actin. The relative level of phosphorylated Drp1 is determined as the ratio of phospho-Drp1 density over actin density (Fig. 2) (*see* **Note 9**).

3.3 Examine Drp1 Tetramer

1. Gently wash the cell with cold PBS twice (*see* **Note 5**).
2. Incubate the cell on ice for 30 min in total cell lysis buffer.
3. Collect the cell and spin for 10 min at 12,000 rpm (or 13,523 × *g*) at 4 °C.
4. Keep the supernatant as total cell lysates.
5. Measure the protein concentration of the total cell lysates and prepare equal amount of proteins for load in each group.
6. Prepare the sample using 1× nonreducing sample loading buffer and boil at 100 °C for 10 min.
7. Run the SDS-PAGE.
8. Transfer the protein to nitrocellulose membrane.
9. Rinse the membrane once with TBST.
10. Block the membrane with blocking solution for 1 h.
11. Cut the membrane into half at the place where the standard maker represents 50 kDa.

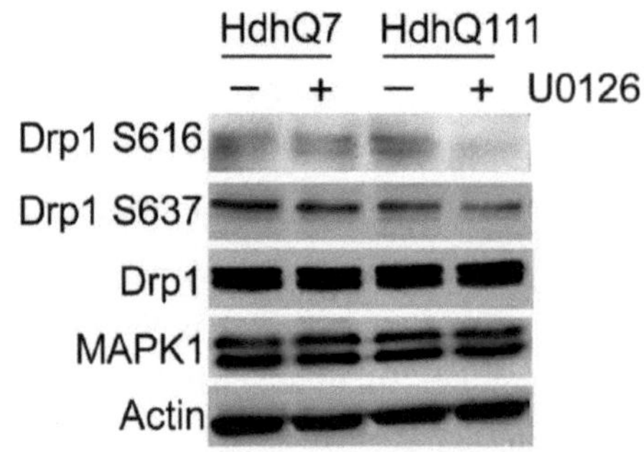

Fig. 2 Drp1 phosphorylation (S616/S637) is examined in Huntington's disease striatal cells (HdhQ7 control, HdhQ111 HD mutant). When Drp1 S637 has no change after the use of MAPK1 inhibitor U0126, the level of Drp1 S616 decreases (Reproduced from [11])

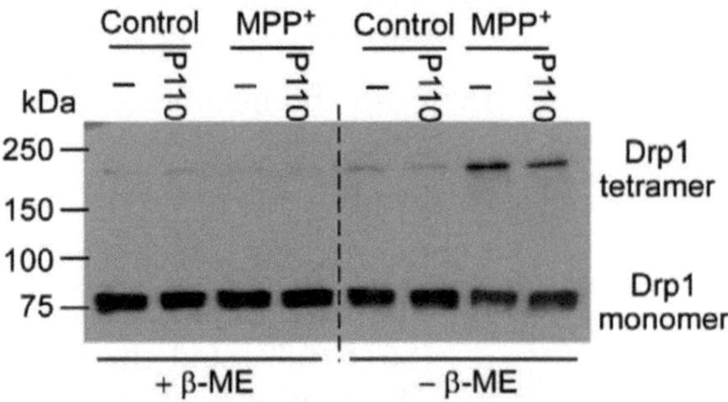

Fig. 3 Drp1 tetramerization can be examined using nonreducing sample loading buffer after the treatment of mitochondrial stressor and is inhibited when using Drp1 inhibitor P110 (Reproduced from Ref. 10)

12. Incubate the top-half membrane with mouse anti-DLP1 antibody diluted in blocking solution (1:1000) and the bottom-half membrane with mouse anti-β-actin diluted in blocking solution (1:1000) at 4 °C overnight.
13. Wash the membranes with TBST for 5 min and totally four times.
14. Incubate the membranes with goat anti-mouse HRP diluted in blocking solution (1:5000) for 1 h.
15. Wash the membrane with TBST for 5 min and totally four times.
16. Incubate the membrane with ECL solution for 2 min.
17. Develop the membrane.
18. The band shown around 75 kDa is Drp1 monomer, whereas the band shown around 200 kDa is Drp1 tetramer. The band shown around 37 kDa is actin. The relative level of Drp1 tetramer is determined as the ratio of Drp1 tetramer density over actin density (Fig. 3).

4 Notes

1. Either self-made or commercially purchased can be used.
2. Protease inhibitor is required for all lysis. Phosphatase inhibitor is required when the lysates are subjected to examining phosphorylated Drp1.
3. Aliquot before freezing.
4. Avoid light. It can be reusable for up to 1 week.
5. PBS wash is not required for tissue samples. For tissue harvest, it is necessary to mince and homogenate the sample when harvesting either mitochondrial fraction or total lysates. The spin conditions are the same as cell culture.
6. Mitochondrial fraction can be stored at −20 °C for 1 week.
7. It is better to check the total Drp1 level as well to exclude the possibility that the change of mitochondrial Drp1 is the result of altered overall Drp1 level.
8. Total cell lysates can be stored at −20 °C for 1 week.
9. It is better to check the total Drp1 level as well to exclude the possibility that the change of phosphorylated Drp1 is the result of altered overall Drp1 level.

Acknowledgment

This work was supported by NIH R01 NS088192 to X.Q.

References

1. Gottlieb RA (2000) Role of mitochondria in apoptosis. Crit Rev Eukaryot Gene Expr 10 (3-4):231–239
2. Jeong SY, Seol DW (2008) The role of mitochondria in apoptosis. BMB Rep 41(1):11–22
3. Chan DC (2006) Mitochondria: dynamic organelles in disease, aging, and development. Cell 125(7):1241–1252
4. Chan DC (2006) Mitochondrial fusion and fission in mammals. Annu Rev Cell Dev Biol **22**:79–99
5. Fannjiang Y et al (2004) Mitochondrial fission proteins regulate programmed cell death in yeast. Genes Dev **18**(22):2785–2797
6. James DI et al (2003) hFis1, a novel component of the mammalian mitochondrial fission machinery. J Biol Chem 278 (38):36373–36379
7. Taguchi N et al (2007) Mitotic phosphorylation of dynamin-related GTPase Drp1 participates in mitochondrial fission. J Biol Chem **282**(15):11521–11529
8. Chang CR, Blackstone C (2007) Cyclic AMP-dependent protein kinase phosphorylation of Drp1 regulates its GTPase activity and mitochondrial morphology. J Biol Chem **282** (30):21583–21587
9. Cribbs JT, Strack S (2007) Reversible phosphorylation of Drp1 by cyclic AMP-dependent protein kinase and calcineurin regulates mitochondrial fission and cell death. EMBO Rep **8**(10):939–944
10. Qi X et al (2013) A novel Drp1 inhibitor diminishes aberrant mitochondrial fission and neurotoxicity. J Cell Sci 126(Pt 3):789–802
11. Roe AJ, Qi X (2018) Drp1 phosphorylation by MAPK1 causes mitochondrial dysfunction in cell culture model of Huntington's disease. Biochem Biophys Res Commun 496 (2):706–711

Chapter 15

Quantifying Drp1-Mediated Mitochondrial Fission by Immunostaining in Fixed Cells

Di Hu and Xin Qi

Abstract

Dynamin-like protein 1 (Drp1) is the master regulator of mitochondrial fission. Drp1 translocates from the cytosol to the mitochondrial outer membrane to execute the scission process. Here we describe an immunofluorescence-based method to measure the mitochondrial translocation of Drp1 and quantify Drp1-related mitochondrial fission by labeling the mitochondrial import receptor subunit TOM20 in fixed cell culture.

Key words Mitochondrial fission, Dynamin-related protein 1, TOM20, Immunofluorescence staining, Confocal microscopy

1 Introduction

Mitochondrial fission is executed by dynamin-related protein 1 (Drp1) along with its adaptor proteins on the mitochondrial outer membrane [1, 2]. Under normal conditions, Drp1 is mainly distributed in cytosol [3]. Various cues including posttranslational modifications (e.g., phosphorylation) and protein–protein interactions, however, induce the mitochondrial translocation of Drp1 to sever mitochondrial filaments at pre-marked constriction sites [3], resulting in the generation of small and rounded mitochondrial fragments. Drp1-mediated mitochondrial fission can be examined by directly visualizing mitochondrial morphology as well as the subcellular localization of Drp1. Here we describe an immunofluorescence method for imaging mitochondria, by targeting the mitochondrial import receptor subunit TOM20, which is widely distributed on mitochondrial outer membrane. Mitochondrial translocation of Drp1 is then analyzed by immunolabeling.

Rajesh Ramachandran (ed.), *Dynamin Superfamily GTPases: Methods and Protocols*, Methods in Molecular Biology, vol. 2159, https://doi.org/10.1007/978-1-0716-0676-6_15,

2 Materials

Prepare all solutions using double-distilled water and analytical grade reagents. Prepare and store all reagents at room temperature (unless indicated otherwise). Follow waste disposal regulations when disposing waste materials.

2.1 Drp1 KO Mouse Embryonic Fibroblast Cell Culture

1. Culture medium: DMEM (Dulbecco's modified Eagle's medium) supplemented with 10% (v/v) heat inactivated FBS and 1% (v/v) penicillin/streptomycin.
2. DPBS (Fisher scientific, Cat. No. SH30028FS).
3. Trypsin (Fisher scientific, Cat. No. SH303236.01).
4. Cell culture plates: 10 cm, 12- and 24-well.

2.2 Coating

1. 12 mm microscope cover glass (Fisher scientific, Cat. No. 12-545-82).
2. 2% Gelatin (Sigma, Cat. No. G1393).

2.3 Transfection

1. Opti-MEM®|reduced-serum medium (Thermo fisher scientific, Cat. No. #31985070).
2. Transfection reagent T2020 (Mirus, Cat. No. MIR5406) (*see* **Note 1**).

2.4 Immuno-fluorescence Staining of Fixed Cell

1. Paraformaldehyde 16% (TED PELLA, Cat. No. 18505).
2. PBS (*see* **Note 2**).
3. Triton X-100 (Fisher scientific, Cat. No. BP-151-500).
4. Normal goat serum (Invitrogen, Cat. No. 10000C).
5. Hoechst 33342 (Invitrogen, Cat. No. H21492).
6. Microscope slides (Fisher scientific, Cat. No. 12-544-2).

2.5 Examine the Expression of Myc-Drp1 by Western Blot

1. Total cell lysis buffer: 50 mM Tris–HCl, pH 7.5, 150 mM NaCl, 1% Triton X-100, and protease inhibitor cocktail.
2. 5% stacking gel.
3. 10% resolving gel.
4. SDS-PAGE running buffer.
5. 2× sample loading buffer (reducing): Add 5 mL 1 M Tris (pH 7), 25 mL 20% SDS, 20 mL glycerol, 2 mg bromophenol blue, 50 μL β-mercaptoethanol, and water up to 100 mL.
6. Nitrocellulose membranes.
7. Transfer buffer.
8. Tris-buffered saline (TBS).

9. TBS containing 0.05% Tween-20 (Anatrace, Cat. No. T1003) (TBST).
10. Blocking solution: 5% milk in TBST. Store at 4 °C.
11. ECL solution: Add 12.2 μL H_2O_2 (10 M), 200 μL luminol (250 mM), 80 μL coumaric acid (90 mM), 4 mL Tris–HCl (1 M, pH 8.5), and water up to 40 mL. Store at 4 °C (*see* **Note 3**).

2.6 Antibodies

1. Mouse anti-c-Myc antibody (Santa Cruz, Cat. No. sc-40).
2. Rabbit anti-Tom20 antibody (Santa Cruz, Cat. No. 11415).
3. Anti-β-Actin antibody (Sigma-Aldrich, Cat. No. A1978).
4. Goat anti-mouse HRP (Thermo Scientific/Fisher, Cat. No. 31430).
5. Alexa 488, goat anti-rabbit IgG (H + L) (Invitrogen, A11034).

3 Methods

Carry out all procedures at room temperature unless otherwise specified. Carry out all of the cell culture studies in cell culture hood. All cells mentioned below are maintained in an incubator at 37 °C in 5% CO_2.

3.1 Restore Drp1 KO Mouse Embryonic Fibroblast

1. Restore a vial of Drp1 knock out (KO) mouse embryonic fibroblast (MEF) [4] from liquid nitrogen or −80 °C.
2. After completely thawing, transfer the cells to a 15 mL tube. Add 10 mL culture medium and gently pipette several times (*see* **Note 4**).
3. Spin down the cells under 1000 rpm (or 400 × *g*) for 5 min.
4. Aspirate the medium and resuspend the cell pellet in 1 mL fresh culture medium. Transfer the cell to a 10 cm cell culture plate with a total of 10 mL medium (*see* **Note 4**) and incubate for 24 h at 37 °C in 5% CO_2.
5. Change the culture medium after 24 h.
6. Split the cells once it reaches 80% confluence.

3.2 Splitting Cells

1. Aspirate the culture medium and wash the cells once with warmed DPBS.
2. Add 1 mL trypsin to the plate and incubate for 3–5 min until most of the cells are detached.
3. Add 5 mL culture medium to the plate and gently pipette down the cells. Transfer the cells to a 15 mL tube (*see* **Note 4**).
4. Spin down the cells at 1000 rpm (or 400 × *g*) for 5 min.

5. Aspirate the medium above the cell pellet in the tube and resuspend the cell pellet in 1 mL culture medium (*see* **Note 4**).
6. Transfer the cells into a new culture plate.

3.3 Coating

1. Put the glass coverslips in 24-well plate (*see* **Note 5**).
2. Sterilize the plate with coverslips under UV inside the cell culture hood for 30 min.
3. Add 500 μL coating solution made of 0.1% gelatin in water in each well to fully cover/merge the glass coverslips.
4. Incubate the 24-well plate under 37 °C for 1 h or under 4 °C overnight.
5. Aspirate out the coating solution and wash with PBS three times before plating the cells on coverslips.
6. Split the Drp1 KO MEF cells and plate 40,000 cells in each well.

3.4 Transfection

1. Transfect cells with empty control vector or Myc-Drp1 vector [5] on the second day following split as detailed below (*see* **Note 1**).
2. For transfection of cells in each well of the 24-well plate, add 50 μL Opti-MEM®|reduced-serum medium to a sterile 1.5 mL Eppendorf tube.
3. Add 0.5 μg of plasmid DNA to the tube.
4. Add 1.5 μL transfection reagent T2020 to the tube and mix by pipetting.
5. Let the tube incubate in the cell culture hood for 25 min.
6. Transfer the medium in the Eppendorf tube to a well (*see* **Note 6**).
7. Cells are ready for immunofluorescence staining 48 h after transfection.

3.5 Immunofluorescence Staining of Mitochondria in Fixed Cell

1. Wash the cells with PBS once.
2. Fix the cell with 4% paraformaldehyde diluted in PBS for 20 min.
3. Wash the cells with PBS three times.
4. Incubate the fixed cells with 0.1% Triton-X-100 diluted in PBS for 5 min.
5. Take out the 0.1% Triton-X-100 and incubate the cells in blocking buffer (2% normal goat serum and 0.05% Triton X-100 in PBS) for 1 h.
6. Take out the blocking buffer and add the primary antibodies, anti-Tom20 and anti-Myc antibodies, diluted in blocking

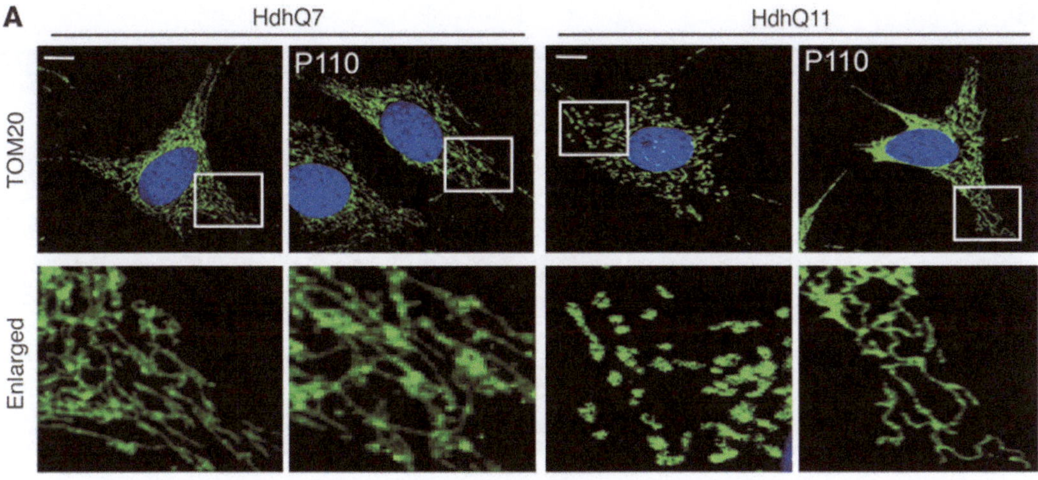

Fig. 1 Treatment with small peptide inhibitor P110 prevented mitochondrial fragmentation in mouse striatal HdhQ111 cells (HdhQ7-control cell, HdhQ111-mutant cell). Tom20 in green. (Reproduced from Guo X et al., Inhibition of mitochondrial fragmentation diminishes Huntington's disease-associated neurodegeneration, J Clin Invest, 123, 5371–5388 (2013))

buffer (1:1000). Incubate overnight at 4 °C or 2 h at room temperature.

7. Wash the cell with PBS three times.
8. Add the secondary antibody, Alexa 488 anti-rabbit IgG and Alexa 568 anti-mouse IgG, diluted in blocking buffer (1:1000). Incubate for 2 h in dark.
9. Wash the cell with PBS three times.
10. Add Hoechst for nuclei staining for 10 min.
11. Wash the cell with PBS three times.
12. Add one droplet of mounting solution on a cover slide. Use tweezers to put the coverslip upside down on the mounting solution, and avoid any bubbles if possible.
13. Dry out the slides overnight at 4 °C in the dark.
14. Use 40× magnification on a fluorescence microscope (e.g., Olympus FV1000 confocal microscope) to observe mitochondrial morphology and Myc-Drp1 distribution the next day as illustrated in Figs. 1 and 2.

3.6 Quantification of Mitochondrial Fragmentation

1. Count the total number of cells immune-positive for anti-Myc antibody, which indicate the Myc-Drp1 expressing cells.
2. Count the total number of Myc-positive cells with fragmented mitochondria among the cells counted in **step 1** as below.

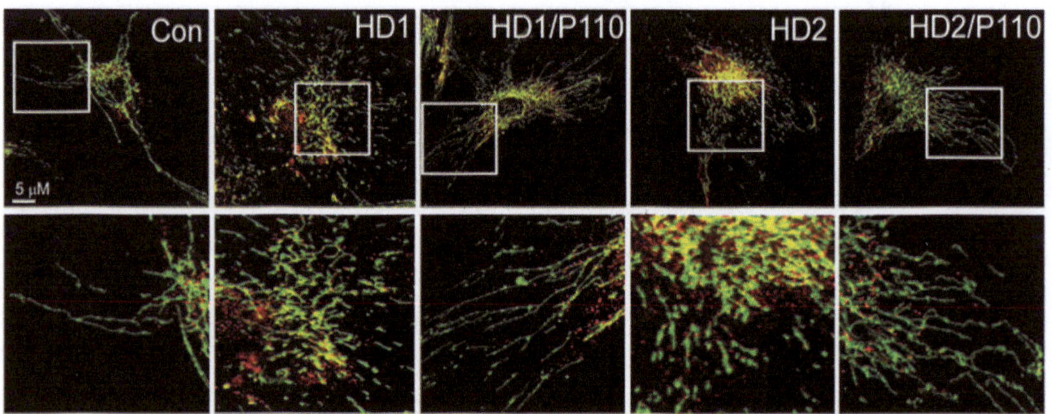

Fig. 2 Specific small peptide inhibitor P110 prevents the mitochondrial translocation of Drp1 and mitochondrial fragmentation in fibroblasts derived from Huntington's disease (HD) patients (Con-control subject, HD1/2-HD patients). Drp1 in red and Tom20 in green. (Reproduced from Guo X et al., Inhibition of mitochondrial fragmentation diminishes Huntington's disease-associated neurodegeneration, J Clin Invest, 123, 5371–5388 (2013))

3. Count at least 100 cells. The percentage of Myc-positive cells with fragmented mitochondrial can be calculated as number of Myc-positive cells with fragmented mitochondria over total number of Myc-positive cells (*see* **Note 7**).

3.7 Examine the Expression Level of Myc-Drp1 in Cells by Western Blot

1. Split the Drp1 KO MEFs into a 12-well plate with 200,000 cells per well.
2. On the second day, transfect the cell with 1 μg Myc-Drp1 vector using 3 μL transfection reagent following the procedures as indicated in Subheading 3.4.
3. After 48 h transfection, gently wash the cells with cold PBS twice.
4. Incubate cells on ice for 30 min in total cell lysis buffer.
5. Collect the cells and spin for 10 min at 12,000 rpm (or 13,523 × *g*) in a table top centrifuge set at 4 °C.
6. Keep the supernatant as total cell lysates.
7. Measure the protein concentration of the total cell lysates and prepare equal amount of proteins for load in each group (*see* **Note 8**).
8. Prepare the sample using 1× reducing sample loading buffer and boil at 100 °C for 10 min.
9. Run SDS-PAGE.
10. Transfer the protein onto nitrocellulose membranes using standard western blotting procedures.

11. Rinse the membrane once with TBST.
12. Block the membrane with blocking solution for 1 h.
13. Cut the membrane at 50 kDa. Incubate the top half of the membrane with anti-c-Myc antibody diluted in blocking solution (1:1000) and bottom half of the membrane with anti-actin antibody at 4 °C overnight.
14. Wash the membranes with TBST for 5 min each for a total of four times.
15. Incubate the membranes with the goat anti-mouse antibody diluted in blocking solution (1:5000) for 1 h.
16. Wash the membrane with TBST for 5 min each for a total of four times.
17. Incubate the membrane with ECL solution for 2 min.
18. Develop the membrane.
19. The band shown around 75 kDa is Myc-Drp1 and the band shown around 37 kDa is actin. The relative level of Drp1 is determined as the ratio of Myc-Drp1 density over actin density.

4 Notes

1. Transfection is not limited to a specific reagent or traditional transfection method.
2. Either self-made or commercially purchased reagents is applicable.
3. Avoid light. Reusable for up to 1 week.
4. Avoid any bubbles.
5. Perform in cell culture hood. It is better to sterilize the tweezer before.
6. It is better to prepare the transfection mixtures as a whole for all of the wells and transfer to each well after 25 min incubation.
7. It is better to average the results of three different coverslips in one experiment as one repeat.
8. 25–35 μg protein in total is enough for each group while preparing the sample for SDS-PAGE.

References

1. Chan DC (2006) Mitochondria: dynamic organelles in disease, aging, and development. Cell 125:1241–1252. https://doi.org/10.1016/j.cell.2006.06.010
2. Chan DC (2006) Mitochondrial fusion and fission in mammals. Annu Rev Cell Dev Biol 22:79–99. https://doi.org/10.1146/annurev.cellbio.22.010305.104638
3. Fannjiang Y, Cheng WC, Lee SJ, Qi B, Pevsner J, McCaffery JM, Hill RB, Basanez G, Hardwick JM (2004) Mitochondrial fission proteins regulate programmed cell death in yeast.

Genes Dev 18:2785–2797. https://doi.org/10.1101/gad.1247904

4. Wakabayashi J, Zhang Z, Wakabayashi N, Tamura Y, Fukaya M, Kensler TW, Iijima M, Sesaki H (2009) The dynamin-related GTPase Drp1 is required for embryonic and brain development in mice. J Cell Biol 186:805–816
5. Macdonald PJ, Stepanyants N, Mehrotra N, Mears JA, Qi X, Sesaki H, Ramachandran R (2014) A dimeric equilibrium intermediate nucleates Drp1 reassembly on mitochondrial membranes for fission. Mol Biol Cell 25:1905–1915

Chapter 16

Imaging Dynamin-Related Protein 1 (Drp1)-Mediated Mitochondrial Fission in Living Cells

Felipe Montecinos-Franjola and Rajesh Ramachandran

Abstract

Mitochondria form highly dynamic networks that continuously undergo fission and fusion. Dynamin-related protein 1 (Drp1), a key regulator of mitochondrial division, self-assembles into a helical polymer around pre-marked scission sites and generates the constriction force necessary to sever the organelle. Live-cell fluorescence imaging of Drp1 oligomerization dynamics and mitochondrial fission can provide unprecedented insights into the spatiotemporal relationship between these coupled processes. The high-resolution images provided by the laser scanning confocal microscope facilitate the observation of the finer details of mitochondrial structure as well as Drp1 polymer dynamics in real time. We provide a detailed description of the confocal imaging methods used to characterize mitochondrial dynamics in living cells with an emphasis on Drp1-mediated mitochondrial fission.

Key words Mitochondrial dynamics, Fission, Drp1, Confocal microscopy, Live-cell imaging

1 Introduction

Mitochondria are double membrane-bound organelles present in all eukaryotic cells [1]. Mitochondria are involved in many essential cellular functions, including ATP production, intracellular calcium signaling, and programmed cell death (apoptosis) [2]. These pivotal functions are intimately connected, however, to mitochondrial morphology.

Mitochondria form highly dynamic networks that continuously undergo regulated cycles of fission and fusion [3]. Depending on the metabolic state of the cell, mitochondria are mostly either tied in highly branched, filamentous, reticular networks or fragmented into discrete, punctate, granular entities spread throughout the cytoplasm [4]. Most often, mitochondria exist as a mixture of both. A dynamic

The original version of this chapter was revised. The correction to this chapter is available at https://doi.org/10.1007/978-1-0716-0676-6_17

Electronic supplementary material: The online version of this chapter (https://doi.org/10.1007/978-1-0716-0676-6_16) contains supplementary material, which is available to authorized users.

Rajesh Ramachandran (ed.), *Dynamin Superfamily GTPases: Methods and Protocols*, Methods in Molecular Biology, vol. 2159, https://doi.org/10.1007/978-1-0716-0676-6_16, © Springer Science+Business Media, LLC, part of Springer Nature 2020

balance between the opposing processes of mitochondrial fission and fusion governs the overall shape of mitochondria in any given cell.

Dynamin superfamily proteins (DSPs) regulate the mitochondrial fission–fusion balance [5, 6]. Drp1 mediates mitochondrial fission and is recruited from the cytosol to the mitochondrial surface at pre-marked fission sites [7]. Membrane-integrated DSPs mitofusins (Mfn) and optic atrophy 1 (OPA1), anchored in the outer and inner mitochondrial membranes, respectively, conversely mediate mitochondrial fusion [5]. In the context of fission, how Drp1 oligomerization dynamics on the mitochondrial membrane surface is coupled to mitochondrial division stills remains poorly understood.

Live-cell confocal microscopy harnessing genetically encoded biomarkers of the mitochondria combined with GFP fusion variants of Drp1 enables correlation of Drp1 membrane dynamics to mitochondrial fission in real time. Here, we detail methods used in the confocal fluorescence imaging of mitochondrial dynamics (Subheading 2) and Drp1-mediated mitochondrial fission (Subheading 3) in living, cultured mammalian cells.

2 Section I: Live-Cell Imaging of Mitochondrial Dynamics

A popular method to visualize the mitochondrial network in living cells is confocal fluorescence microscopy that allows for real-time monitoring of mitochondrial dynamics at high resolution and low noise. Specific labeling of the mitochondria can be achieved either with the use of synthetic fluorophores or with genetically encoded biomarkers. A common synthetic dye for staining mitochondria is MitoTracker that offers the advantage of covalent attachment to the mitochondria and is insensitive to cell fixation and/or death [8]. However, many synthetic fluorescent dyes can be toxic to cells even when illuminated for short periods of time (phototoxicity) [9] and are susceptible to rapid photobleaching. Genetically encoded biosensors, such as Mito-mCherry composed of a mitochondrial targeting sequence (mito) fused to the N-terminus of the monomeric red fluorescent protein (mCherry) [10] (Fig. 1), on the other hand, allow imaging of the mitochondria under controlled expression conditions and are generally not phototoxic.

The confocal fluorescence microscope allows acquisition of high-resolution images in real time. Post-processing of the images with specialized software, such as ImageJ [11], can be used to extract both qualitative and quantitative information about mitochondrial morphology and dynamics. With the application of slice acquisition through z-stacks in the confocal microscope, the researcher can now obtain improved volumetric information by 3D reconstruction of subcellular structures such as the mitochondrial network [12]. This section describes the steps for sample preparation, acquisition of high-resolution images, and post-acquisition image processing for time-lapse microscopy and 3D reconstruction of the mitochondrial network.

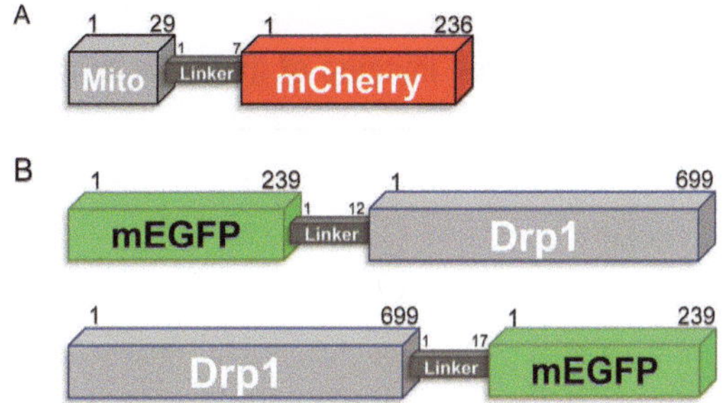

Fig. 1 Schematic representation of the fluorescent protein constructs used for live-cell imaging. (**a**) mCherry-Mito-7 (Addgene #55102) comprised of the mitochondrial targeting sequence of subunit VIII of human cytochrome C oxidase (29 aa) fused to the N-terminus of mCherry (236 aa, accession AY678264). (**b**) Human Drp1 (699 aa isoform 3, accession NP_005681) fused to monomeric enhanced green fluorescent protein (mEGFP carrying A206K mutation, 239 aa, accession AF323988_1) either at the N- or C-terminus, subcloned in plasmids pEGFP-C1 and pEGFP-N1 (Clontech), respectively

2.1 Materials

1. Incubator with temperature and humidity control, and a CO_2-controlled atmosphere.
2. Cells (HeLa or MEFs) stably or transiently expressing any genetically encoded mitochondrial biomarker, such as mCherry-Mito-7 (Addgene #55102; Fig. 1), cultured in a suitable growth medium. Standard transfection protocols (e.g., lipofectamine-based) are better suited for this purpose.
3. Transparent cultivation dishes—35 mm diameter, #1.5 cover glass (0.16–0.19 mm thickness) (MatTek).
4. Laser scanning microscope system (e.g., Olympus FV1000) equipped with a heated stage. 40× or 60× oil or water immersion objective with high NA (>1.2). Laser lines appropriate for excitation of the fluorescent protein of interest. Appropriate filter cubes and dichroic mirrors are normally included in the microscope.
5. Cell culture medium for imaging must provide nutrients essential for cellular health and should be transparent to avoid excessive background signal during imaging. Examples are DPBS, DMEM without phenol red, or FluoroBrite DMEM Media (ThermoFisher).
6. Computer with image processing software such as Fiji or ImageJ (free and open source) [11].

2.2 Methods

1. Plate the cells in a transparent cultivation chamber with suitable growth medium such as DMEM supplemented with 10% fetal bovine serum (FBS) and 1% antibiotics (e.g., penicillin and streptomycin). Incubate at 37°C and 5% CO_2 (*see* **Note 1** for details on the handling of frozen cell stocks).

Table 1
Acquisition parameters for the imaging of live cells using the confocal microscope

Parameter[a]	Value
Excitation wavelength (nm)	488 (EGFP), 543 (mCherry)
Laser intensity (%)	2–10
Pixel time, μs	4–12
Frame size	512 × 512 or 1024 × 1024
Digital zoom	1–5
Pinhole (confocal aperture), Airy units	0.8–1.1
Detector settings	Photon counting (recommended) or analog
Step size, μm (z-stack)	0.1–0.5

[a]Parameters optimized for acquisition on the Olympus FV1000 microscope

2. For transient expression of the protein of interest (e.g., mCherry-Mito-7), cells should be incubated for 24–48 h post-transfection to allow expression and maturation of the biosensor. Stably expressing cell lines can be transferred to a suitable transparent cultivation chamber for imaging. However, incubation for 24 h prior to imaging is recommended.
3. Replace the culture medium with transparent imaging medium such as DPBS or DMEM without phenol red (*see* **Notes 2** and **3** for tips on changing cell culture medium with imaging medium). Incubate the cell for 1–2 h to enable the cells to adjust to the new solution.
4. Turn on the microscope system and preheat the microscope stage to 37 °C. Allow the system to equilibrate to environmental conditions and let the laser source stabilize for at least 30 min prior to the experiment.
5. Set the instrument and acquisition software parameters prior to setting up the sample in the microscope stage (*see* **Note 4** for tips on setting up the microscope optics for these experiments). Refer to Table 1 for a list of parameters optimized for image acquisition in the Olympus FV1000 laser scanning microscope.
6. Put a drop of oil or water on the objective (60×) and properly clamp the imaging chamber onto the stage to avoid any movements during image acquisition (*see* **Note 5** for recommendations on handling the microscope and the sample holder).
7. Transfer the dish with cells from the incubator to the microscope stage preset at the working temperature. If a controlled atmosphere stage (i.e., temperature-, humidity-, and CO_2-controlled) for the microscope is not available, we suggest using the cells for no more than 2 h (*see* also **Note 6**).

8. Once the microscope system is set with the sample chamber securely attached to the stage, let the system adjust to the temperature and atmospheric conditions for 30 min.
9. Using the microscope eyepiece, epifluorescence, and focusing tools (e.g., hand knobs), select a group of cells showing good expression of the fluorescent protein. In the case of mCherry-Mito-7, select an appropriate fluorescence filter cube (e.g., Cy3/TRITC) for imaging. This step is critical as it helps focus the objective lens on the region of interest (ROI) prior to confocal image acquisition.
10. Once the cells are selected and the focal distance adjusted, you may switch to laser scanning mode and use the FAST scan to obtain a raw image of the fluorescent cells. Start the image acquisition with low laser power (~2%; *see* Table 1 for parameters) and short pixel time (4–8 μs) to avoid photodamage and minimize photobleaching (*see* **Note** 7 for tips on adjusting the detector settings).
11. If the cells are in good shape and the fluorescence does not saturate the detector, then you can switch to optimal acquisition parameters (Table 1) and collect the high-resolution images of the fluorescent cell. An example of a low expressing cell of the fusion protein mCherry-Mito-7 that stains the mitochondrial network is shown in Fig. 2a. The image was acquired in 8 s using 1024 × 1024 frame size, 8 μs pixel time and a zoom of 3.
12. For time-lapse imaging, input the number of images (frames) to be collected and set the instrument to scan. Two options are considered. Either choose time scan by setting a scanning frequency (e.g., every 5 s) or simply select the number of frames desired (e.g., 100 frames). The system will automatically calculate the time required to collect such number of images as long as the acquisition parameters are set as previously noted (like frame size and pixel time). If focal drift is a problem, use autofocus device if present (*see* **Note 8** for more details).
13. A montage of images collected using time-lapse acquisition is shown in Fig. 2b (*see* also Supplementary Movie 1). The time-lapse shows the event of mitochondrial transport across the cytoplasm, followed by the fission of a mitochondrial fragment, and the subsequent fusion of one of the resulting daughter mitochondria with a preexisting and stable mitochondrial filament.
14. Complementary information about changes in mitochondrial morphology is obtained by the use of z-stack series and post-acquisition image processing for 3D reconstruction. The hyperfused mitochondrial network of *Drp1*$^{-/-}$ MEFs

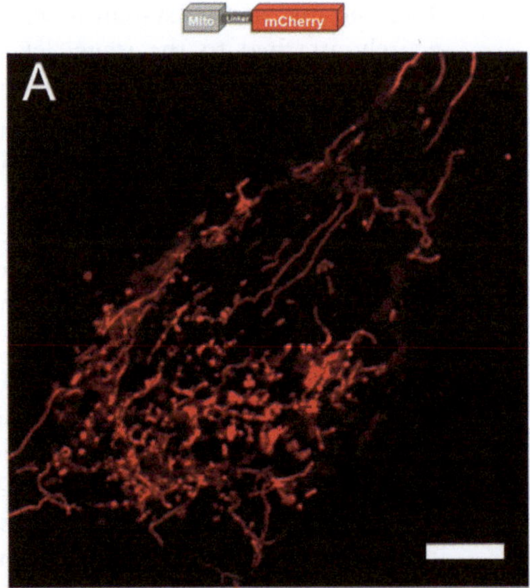

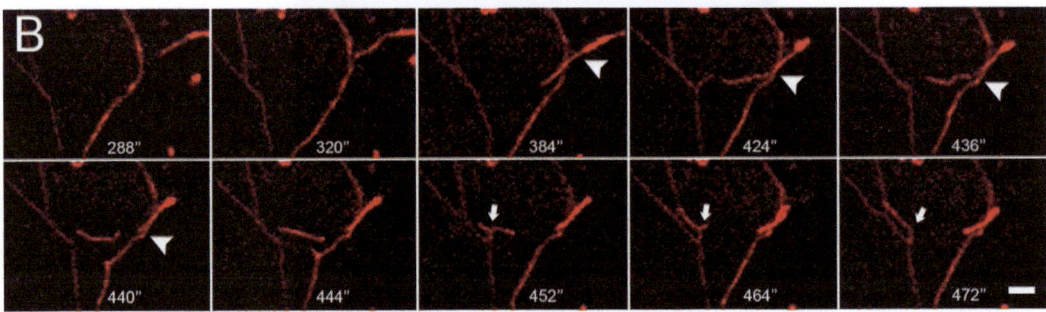

Fig. 2 Live imaging of mitochondrial network dynamics. (**a**) High-resolution confocal imaging of the mitochondrial network in live HeLa cells transiently expressing the fusion protein mCherry-Mito-7. (**b**) Real-time observation of mitochondrial fission and fusion. mCherry was excited using the 543 nm laser line, and 512 × 512 images were recorded every 4 s. A white arrowhead indicates the fission of a growing mitochondrial filament, whereas a white arrow indicates the subsequent fusion of one of the daughter fragments with another mitochondrial filament nearby. Scale bar, panel **a**, 10 μm. Scale bar, panel **b**, 4 μm. *See* Supplementary Movie 1 for a more detailed view

[13, 14] is shown in Fig. 3. The z-stack collected every 0.3 μm (all other parameters of acquisition are noted in Table 1) is shown as a montage in Fig. 3a. The maximum intensity projection that combines all the slices in a 2D representation is shown in Fig. 3b. The resulting 3D reconstruction of the mitochondria is represented as volume in Fig. 3c. *See* also Supplementary Movies 2 and 3 for animated versions of the z-stack and of the 3D reconstruction, respectively (*see* **Note 9** for more details on the processing of these images).

2.3 Notes

1. The cells should be subject to two to three passages after thawing from frozen stocks. The cells are preferably thawed quickly in a 37 °C water bath and immediately diluted in

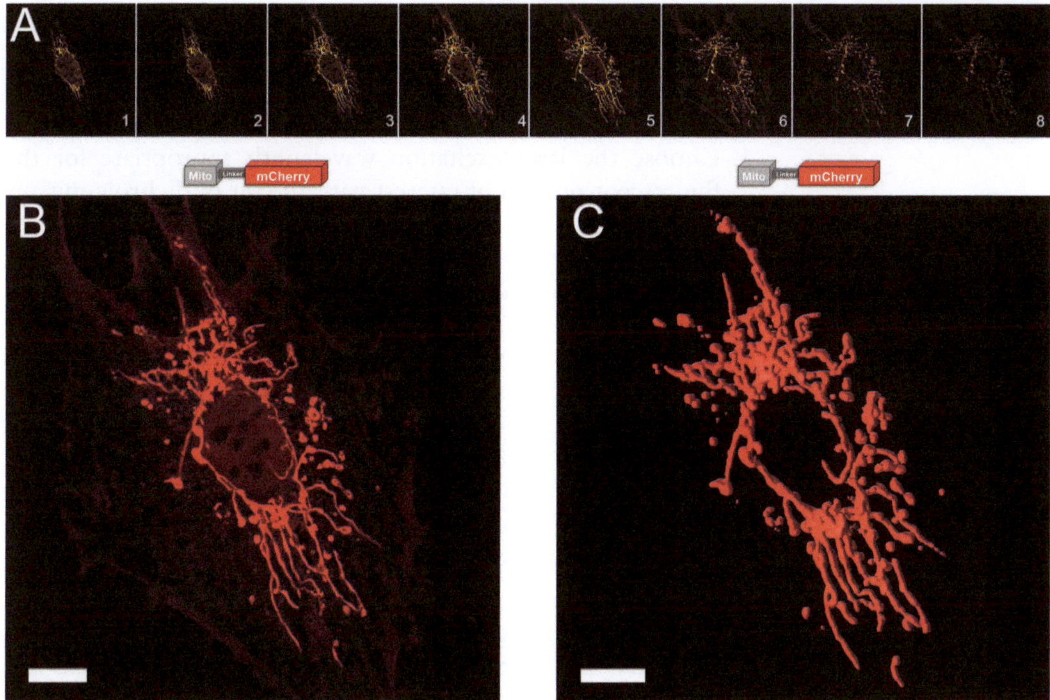

Fig. 3 Live-cell imaging of the hyperfused mitochondrial network in *Drp1*$^{-/-}$ MEFs transiently expressing the fusion protein mCherry-Mito-7 (red). (**a**) Montage of the 512 × 512 z-stack series acquired every 0.3 μm with the confocal microscope (*see* also Supplementary Movies 1 and 3). (**b**) Maximum intensity projection of the z-stack. (**c**) 3D reconstruction of the mitochondrial network. The mitochondria appear elongated (or hyperfused) due to lack of fission in the absence of Drp1. Scale bar, 10 μm

pre-warmed growth medium before seeding in the incubation dish. Multiple passages of the revived cells from frozen stocks enable the cells to maintain viability, given that the passages are made before they reach confluency (i.e., in log phase).

2. The culture medium for growing cells can be used for imaging, but it has the caveat of containing components, such as the pH indicator phenol red, that are highly fluorescent and contribute to the background noise. High background fluorescence may mask the signals of weakly expressing fluorescent biosensors. To replace this medium, use transparent alternatives that are not fluorescent and support normal cell growth when supplemented with the appropriate nutrients. When changing the medium, try washing once or twice to remove traces of the normal DMEM. We prefer using transparent medium specifically designed for imaging such as FluoroBrite (ThermoFisher) or Live Cell Imaging Solution (Invitrogen) that can maintain healthy cells for up to 4 h.

3. Aseptic technique to reduce the potential contamination of the medium with unwanted microorganisms is key for the

appropriate handling of cell cultures. This is especially important when frequently changing the medium. Use sterile solutions pre-warmed at the working temperature and handle all materials in a laminar flow hood.

4. Choose the laser excitation wavelength appropriate for the fluorescent protein of interest with the set of dichroic mirrors and band-pass filters to select the fluorescence signals accordingly. For acquisition of mEGFP and mCherry fluorescence, we use 488 and 543 nm excitation, respectively. For dual excitation, the fluorescence is collected through a multiband dichroic mirror (e.g., DM405/488/543/635 for the Olympus FV1000). Multichannel detection allows the separation of the fluorescent signals from different fluorophores (e.g., using band-pass filters BA505-525 for mEGFP and BA560-660 for mCherry). As a rule, use low excitation power to avoid photodamage due to the heat generated and to minimize photobleaching in long-term acquisition experiments.
5. The optics and stability of the microscope components are very sensitive to environmental variations such as temperature, humidity, and airflow. It is recommended to maintain the microscope in a closed dark room (to avoid parasitic light entering the detector) with sufficient airflow to permit long-term stability of the system. Avoid placing the microscope system close to an air ventilation vent, as the airflow will affect the focus and temperature stability. Use a small fan to improve airflow if no other option is available.
6. Ideally, the culture dish containing the cells is placed in an atmosphere-controlled stage in the microscope. However, if such device is not available, the acquisition of images should be limited to the time span in which the cells are healthy. Timing is very important as the morphology of the cell as well as the mitochondrial network responds to changes in solution parameters such as pH, temperature, concentration of nutrients, etc. As a general rule, the cells should not be imaged continuously for more than 4 h under such conditions. If controlled atmosphere is available, with the appropriate imaging medium, the cells can be imaged continuously for up to 2 h, provided that photobleaching is not a limiting factor.
7. When selecting the detector settings in the Olympus FV1000, there are two options for mode: photon counting or analog. The photon-counting mode is preferred because it provides a better signal-to-noise ratio and a better filtering of noise from the signal (good for low signals). The analog mode is more popular but has the caveat of not discriminating background noise from the signal. The noise is added to the overall signal and cannot be removed. However, the analog mode can be

useful when dealing with high signals that may saturate the PMT in the photon-counting mode, as detector sensitivity can be adjusted in the analog mode (in photon-counting mode, the sensitivity is set to maximum and cannot be adjusted). The choice of one mode over the other is dependent on the experimental needs. In the photon-counting mode, adjust the excitation power to obtain a good signal.

8. Time-lapse imaging requires careful consideration of the scanning parameters: laser power, scanning frequency, and stability of the focus. Laser power should be chosen such that sufficient counts are recovered while minimizing the photobleaching. Use low laser power and set the detector to photon-counting mode because it uses maximal sensitivity. Scanning frequency is also key for minimizing photobleaching, while enabling the capture of the time-dependent mitochondrial movements in a live cell. Under normal circumstances, since the kinetics of mitochondrial dynamics are moderate to fast at 37 °C, we collect images every 4–5 s. The focus stability is crucial in time-lapse imaging to enable imaging of the organelle for long periods. Some microscope systems contain a focal drift compensation device that corrects for loss of focus caused by temperature changes, cellular movements, and other factors. Autofocus is recommended for long-term time-lapse imaging.

9. In the acquisition of z-stack, a high laser power should be avoided because photobleaching will have a big impact on the quality of the slices used for 3D reconstruction. The stability of the focus is also very important since the movements are in the sub-micrometer range. Suggestions and details about the settings and parameters for data acquisition and 3D reconstruction are provided in Refs. 12, 15. Once the images are collected and saved, they can be directly loaded in ImageJ for processing. The z-stack is then loaded with the plugin "3D viewer," and by using the default settings, it is possible to obtain the 3D reconstruction shown in Fig. 3. However, some adjustments are required. With z-stack series loaded in ImageJ, go to tab "Image" and select "properties." A window will pop-up where you can enter the units of length (obtained from the microscope information files), pixel width, height, and depth. If the appropriate parameters are entered, then the 3D reconstruction will have the correct spatial information. It is important to be careful with interpreting lengths and sizes in general since the radial and axial resolution of the confocal can be difficult to ascertain.

3 Section II: Imaging Drp1-Mediated Mitochondrial Fission in Live Cells

To investigate the role of Drp1 dynamics in mitochondrial fission, we use the genetically encoded fusion of Drp1 to monomeric EGFP (Fig. 1b). Both N- and C-terminal fusions of Drp1 to GFP have been described in the literature [16, 17]. Either construct localizes to discrete puncta on the mitochondrial surface. The Drp1 puncta generally appear evenly distributed along the length of the networked mitochondrial tubules, and time-lapse imaging shows that these puncta co-localize with sites of mitochondrial fission [7]. In this section, we will describe the steps necessary for obtaining high-quality images of live cells expressing EGFP fusion constructs of Drp1 in combination with mitochondrial staining using mCherry-Mito-7. We also demonstrate the use of various tools to track the events of Drp1-mediated mitochondrial fission in real time.

3.1 Materials

1. The materials required for these experiments are the same as detailed in Subheading 2.1 for the fluorescence staining of mitochondria.
2. The expression plasmids for monomeric EGFP fusion with Drp1 (N- or C-terminally tagged; Fig. 1) are delivered by co-transfection with mCherry-Mito-7 expressing plasmid (Subheading 2.1), and the cell culture is handled as explained above.

3.2 Methods

1. Repeat **steps 1** through **8** of Subheading 2.2. However, there are considerations which may help in the interpretation of the imaging data that relate to the expression level of the protein constructs (*see* **Note 1** of Subheading 3.3 for some recommendations).
2. Selection of the cells for imaging should be done with eyepiece of the microscope. Select for those cells showing low expression levels but with sufficient signal for easy detection. High-expressing cells display gross Drp1 aggregation in the cytoplasm (a problem associated with GFP tagging) as well as altered mitochondrial morphology.
3. Set up the microscope acquisition parameters as described in steps 9 through 12 of Subheading 2.2. Here, for cells co-transfected with mCherry-Mito-7 and either mEGFP-Drp1 or Drp1-mEGFP, the excitation of the sample is performed with both 488 and 543 nm lasers simultaneously (*see* **Note 4** of Subheading 2.3 for more details about the microscope acquisition settings).
4. The localization of the fluorescent Drp1 constructs in the cytoplasm and over mitochondria in HeLa cells is shown in Fig. 4a, b. In these cells, the mitochondrial network appears

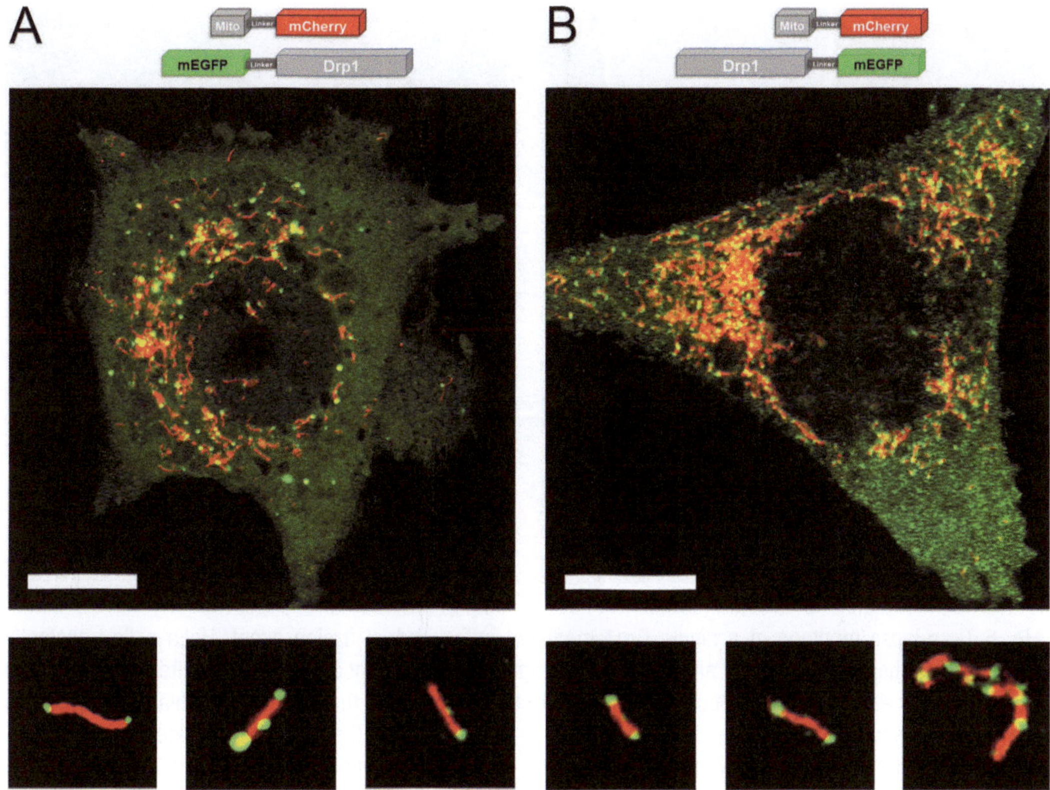

Fig. 4 Live-cell confocal imaging of HeLa cells expressing mCherry-Mito-7 along with either mEGFP-Drp1 (panel **a**) or Drp1-mEGFP (panel **b**). EGFP (green) and mCherry (red) were excited with 488 and 543 nm laser lines, respectively. 1024 × 1024 images were acquired. Drp1 is predominantly localized in the cytoplasm, but form discrete puncta (higher order oligomers) over mitochondria. Mitochondria appear hyperfragmented due to overexpression of the mitochondrial fission DSP, Drp1. Zoomed images at the bottom show mitochondrial fission products typically observed in these cells, with Drp1 localized either at the poles or at imminent division sites. Scale bar 10 μm

hyperfragmented due to Drp1 overexpression [16, 17]. Drp1 localization in discrete puncta over mitochondria is clearly observed. Drp1 localizes to sites of future fission, with puncta evenly distributed along the length of the mitochondrial filaments. Drp1 is also visualized at the poles of short mitochondrial fragments newly generated by fission (lower panels in Fig. 4). These images show the expected distribution of the Drp1 fluorescent constructs indicating that the fusion proteins retain function.

5. The expression of mEGFP-Drp1 along with mCherry-Mito-7 can be used to specifically follow the dynamics of Drp1-mediated mitochondrial fission. Figure 5 shows a montage of time-lapse images demonstrating Drp1-mediated mitochondrial fission. The lariat-shaped mitochondrial fragment displays a single Drp1 punctum and undergoes fission at the exact site

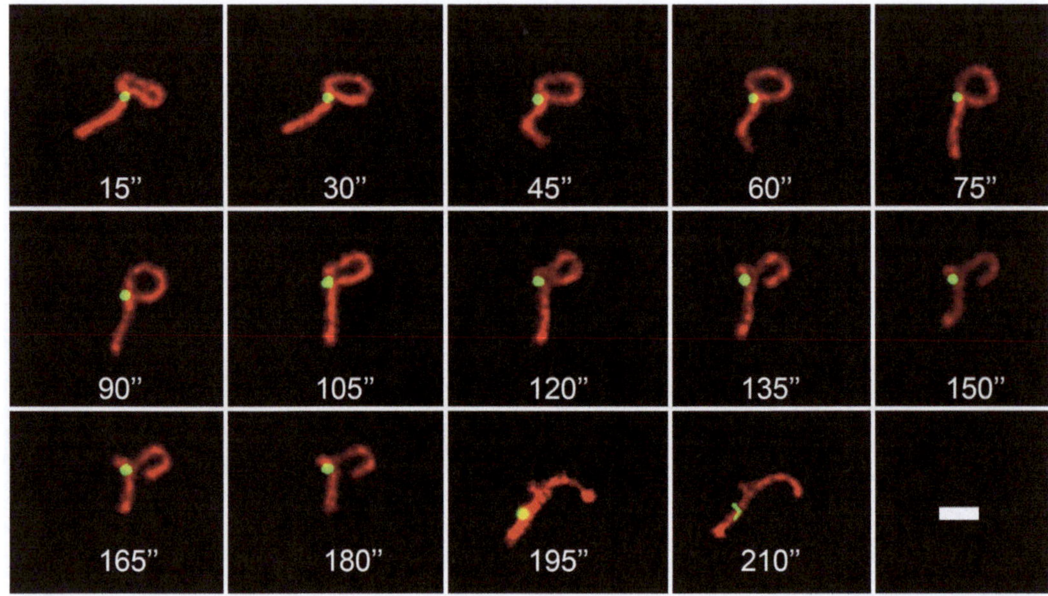

Fig. 5 Time-lapse montage of a single Drp1-mediated mitochondrial fission event. HeLa cells transiently expressing mCherry-Mito-7 (red) and mEGFP-Drp1 (green) were imaged every 5 s excited with 488 and 543 nm laser lines, respectively. Starting with a lariat-shaped structure, the mitochondrial filament (red) binds Drp1 (bright green puncta), which mediates the fission and opening of the lariat into an extended mitochondrial fragment. Partial disassembly of the Drp1 polymer (decreased brightness) post-fission could be observed in the last frame. Scale bar, 4 μm. *See* Supplementary Movie 4 for an animated version of the fission event

of Drp1 localization. *See* also Supplementary Movie 4 for an animated version of the fission event. These experiments can be further used for quantitation of mitochondrial fission kinetics.

3.3 Note

1. Transfection has the advantage of providing a relatively easy way to deliver DNA into cells for expression of proteins of interest. However, there are some caveats that need consideration. For example, using large amounts of DNA in the transfection reaction (i.e., greater than 2 μg DNA) can lead to a very high expression of the protein of interest, especially at high transfection efficiency. Appropriate controls should be carried out to test the effect of the expression level over cell viability. In the case of mitochondrial staining, it is important to ascertain that mitochondrial morphology is not affected by the expression of mCherry-Mito-7. Compare morphology with the mitochondria of fixed cells revealed by immunostaining (i.e., without transfection). Another option is to characterize cell viability by comparison with a cell transfected with the empty vector. If the transfection is found to affect mitochondrial morphology or the viability of cells, then consideration should be given to develop a stable cell line.

Acknowledgments

The cell lines used in this study were obtained from the labs of Ting-wei Mu (HeLa) and Xin Qi (*Drp1*$^{-/-}$ MEFs), both of Case Western Reserve University School of Medicine. FM-F is grateful to Yanlin Fu and Di Hu (CWRU) for advice on the handling of cell cultures and co-transfection. This work was supported by National Institutes of Health grant R01GM121583 awarded to R.R.

References

1. Friedman JR, Nunnari J (2014) Mitochondrial form and function. Nature 505:335–343
2. Youle RJ, van der Bliek AM (2012) Mitochondrial fission, fusion, and stress. Science 337:1062–1065
3. Labbe K, Murley A, Nunnari J (2014) Determinants and functions of mitochondrial behavior. Annu Rev Cell Dev Biol 30:357–391
4. Mishra P, Chan DC (2016) Metabolic regulation of mitochondrial dynamics. J Cell Biol 212:379–387
5. Ramachandran R (2018) Mitochondrial dynamics: the dynamin superfamily and execution by collusion. Semin Cell Dev Biol 76:201–212
6. Ramachandran R, Schmid SL (2018) The dynamin superfamily. Curr Biol 28: R411–R416
7. Friedman JR, Lackner LL, West M, DiBenedetto JR, Nunnari J, Voeltz GK (2011) ER tubules mark sites of mitochondrial division. Science 334:358–362
8. Chazotte B (2009) Labeling mitochondria with fluorescent dyes for imaging. Cold Spring Harb Protoc 2009:pdb prot4948
9. Minamikawa T, Sriratana A, Williams DA, Bowser DN, Hill JS, Nagley P (1999) Chloromethyl-X-rosamine (MitoTracker Red) photosensitises mitochondria and induces apoptosis in intact human cells. J Cell Sci 112 (Pt 14):2419–2430
10. Olenych SG, Claxton NS, Ottenberg GK, Davidson MW (2007) The fluorescent protein color palette. Curr Protoc Cell Biol 21:25
11. Schindelin J, Arganda-Carreras I, Frise E, Kaynig V, Longair M, Pietzsch T, Preibisch S, Rueden C, Saalfeld S, Schmid B, Tinevez JY, White DJ, Hartenstein V, Eliceiri K, Tomancak P, Cardona A (2012) Fiji: an open-source platform for biological-image analysis. Nat Methods 9:676–682
12. Simula L, Campello S (2018) Monitoring the mitochondrial dynamics in mammalian cells. Methods Mol Biol 1782:267–285
13. Wakabayashi J, Zhang Z, Wakabayashi N, Tamura Y, Fukaya M, Kensler TW, Iijima M, Sesaki H (2009) The dynamin-related GTPase Drp1 is required for embryonic and brain development in mice. J Cell Biol 186:805–816
14. Macdonald PJ, Stepanyants N, Mehrotra N, Mears JA, Qi X, Sesaki H, Ramachandran R (2014) A dimeric equilibrium intermediate nucleates Drp1 reassembly on mitochondrial membranes for fission. Mol Biol Cell 25:1905–1915
15. Mitra K, Lippincott-Schwartz J (2010) Analysis of mitochondrial dynamics and functions using imaging approaches. Curr Protoc Cell Biol Chapter 4:Unit 4 25 21–Unit 4 25 21
16. Smirnova E, Griparic L, Shurland DL, van der Bliek AM (2001) Dynamin-related protein Drp1 is required for mitochondrial division in mammalian cells. Mol Biol Cell 12:2245–2256
17. Strack S, Cribbs JT (2012) Allosteric modulation of Drp1 mechanoenzyme assembly and mitochondrial fission by the variable domain. J Biol Chem 287:10990–11001

Correction to: Imaging Dynamin-Related Protein 1 (Drp1)-Mediated Mitochondrial Fission in Living Cells

Felipe Montecinos-Franjola and Rajesh Ramachandran

Correction to:
Chapter 16 in: Rajesh Ramachandran (ed.), *Dynamin Superfamily GTPases: Methods and Protocols*, Methods in Molecular Biology, vol. 2159,
https://doi.org/10.1007/978-1-0716-0676-6_16

The chapter was inadvertently published with the incorrect unit mm (millimeters) as "Step size, mm (z-stack)" instead of μm (micrometers) as "Step size, μm (z-stack)".

The correction has been incorporated by changing the units from "mm" (millimeters) to "μm" (micrometers) and has been updated as "Step size, μm (z-stack)".

Table 1
Acquisition parameters for the imaging of live cells using the confocal microscope

Parameter[a]	Value
Excitation wavelength (nm)	488 (EGFP), 543 (mCherry)
Laser intensity (%)	2–10
Pixel time, μs	4–12
Frame size	512 × 512 or 1024 × 1024
Digital zoom	1–5
Pinhole (confocal aperture), Airy units	0.8–1.1
Detector settings	Photon counting (recommended) or analog
Step size, μm (z-stack)	0.1–0.5

[a]Parameters optimized for acquisition on the Olympus FV1000 microscope

The updated online version of the chapter can be found at
https://doi.org/10.1007/978-1-0716-0676-6_16

Rajesh Ramachandran (ed.), *Dynamin Superfamily GTPases: Methods and Protocols*, Methods in Molecular Biology, vol. 2159,
https://doi.org/10.1007/978-1-0716-0676-6_17,

Index

A

B

C

D

E

F

Rajesh Ramachandran (ed.), *Dynamin Superfamily GTPases: Methods and Protocols*, Methods in Molecular Biology, vol. 2159, https://doi.org/10.1007/978-1-0716-0676-6, © Springer Science+Business Media, LLC, part of Springer Nature 2020

MIX
Papier aus verantwortungsvollen Quellen
Paper from responsible sources
FSC® C105338

If you have any concerns about our products,
you can contact us on
ProductSafety@springernature.com

In case Publisher is established outside the EU,
the EU authorized representative is:
Springer Nature Customer Service Center GmbH
Europaplatz 3, 69115 Heidelberg, Germany

Printed by Libri Plureos GmbH
in Hamburg, Germany